Patrick Moore's Practical Astronomy Series

Springer
London
Berlin
Heidelberg
New York
Hong Kong
Milan
Paris
Tokyo

Other titles in this series

Telescopes and Techniques (2nd Edn.)
Chris Kitchin

The Art and Science of CCD Astronomy
David Ratledge (Ed.)

The Observer's Year
Patrick Moore

Seeing Stars
Chris Kitchin and Robert W. Forrest

Photo-guide to the Constellations
Chris Kitchin

The Sun in Eclipse
Michael Maunder and Patrick Moore

Software and Data for Practical Astronomers
David Ratledge

Amateur Telescope Making
Stephen F. Tonkin (Ed.)

Observing Meteors, Comets, Supernovae and other Transient Phenomena
Neil Bone

Astronomical Equipment for Amateurs
Martin Mobberley

Transit: When Planets Cross the Sun
Michael Maunder and Patrick Moore

Practical Astrophotography
Jeffrey R. Charles

Observing the Moon
Peter T. Wlasuk

Deep-Sky Observing
Steven R. Coe

AstroFAQs
Stephen Tonkin

The Deep-Sky Observer's Year
Grant Privett and Paul Parsons

Field Guide to the Deep Sky Objects
Mike Inglis

Choosing and Using a Schmidt-Cassegrain Telescope
Rod Mollise

Astronomy with Small Telescopes
Stephen F. Tonkin (Ed.)

Solar Observing Techniques
Chris Kitchin

Observing the Planets
Peter T. Wlasuk

Light Pollution
Bob Mizon

Using the Meade ETX
Mike Weasner

Practical Amateur Spectroscopy
Stephen F. Tonkin (Ed.)

More Small Astronomical Observatories
Patrick Moore (Ed.)

Observer's Guide to Stellar Evolution
Mike Inglis

How to Observe the Sun Safely
Lee Macdonald

The Practical Astronomer's Deep-Sky Companion
Jess K. Gilmour

Observing Comets
Nick James and Gerald North

Observing Variable Stars
Gerry A. Good

Visual Astronomy in the Suburbs
Antony Cooke

Astronomy of the Milky Way: The Observer's Guide to the Northern and Southern Milky Way (2 volumes)
Mike Inglis

The NexStar User's Guide
Michael W. Swanson

Observing Binary and Double Stars
Bob Argyle (Ed.)

Navigating the Night Sky
Guilherme de Almeida

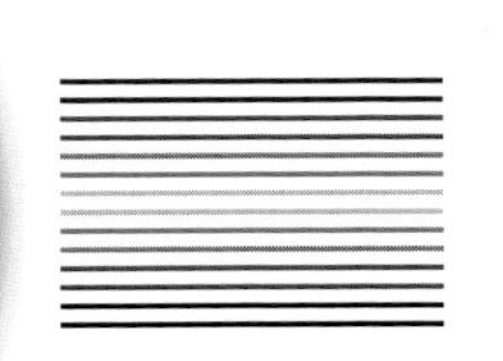

Martin Mobberley

With 124 Figures

Cover illustration: Background: The Theophilus region of the Moon imaged by Mike Brown of York, England (see Chapter 6).

British Library Cataloguing in Publication Data
Mobberley, Martin, 1958–
The new amateur astronomer.–(Patrick Moore's practical astronomy series)
1. Astronomy–Amateurs' manuals 2. Astronomical instruments –Amateurs' manuals 3. Astronomers
I. Title
520
ISBN 1852336633

Library of Congress Cataloging-in-Publication Data
Mobberley, Martin, 1958–
The new amateur astronomer / Martin Mobberley.
p. cm. – (Patrick Moore's practical astronomy series, ISSN 1617-7185)
Includes bibliographical references and index.
ISBN 1-85233-663-3 (pbk. : alk. paper)
1. Astronomy–Amateur's manuals. I. Title. II. Series.
QB63.M595 2004
520–dc22 2004041728

Patrick Moore's Practical Astronomy Series ISSN 1617-7185
ISBN 1-85233-663-3 Springer-Verlag London Berlin Heidelberg
Springer-Verlag is part of Springer Science+Business Media
springeronline.com

Printed in the United States of America

Typeset by EXPO Holdings, Malaysia
58/3830-543210 Printed on acid-free paper SPIN 10880729

Preface

It was a book that initially inspired me to observe the night sky: Patrick Moore's *Observer's Book of Astronomy* (1968 edition). The things that really caught my eye in that book were the superb planetary paintings by the late Leslie Ball and pictures of 1960s amateur telescopes: Henry Brinton's $12\frac{1}{2}$-inch reflector, Patrick's $12\frac{1}{2}$-inch reflector and run-off shed, and J. Hedley Robinson's observatory at Teignmouth.

I was ten at the time, but I knew that one day I wanted to get planetary images like those sketched by Leslie Ball, wanted to own a good telescope, and wanted to write a few books like Patrick. Thirty-five years on, my ambitions have not changed at all, but technology has moved on. Amateurs are now achieving what professionals achieved in 1968, both in terms of limiting magnitude, resolution, and discoveries! The "new" amateur astronomer still strives for the best telescope in his or her backyard, but can now observe in the warm, using CCDs and robotic telescopes. My feelings about that first astronomy book have stayed with me, and I firmly believe that amateurs are still principally inspired by spectacular planetary, cometary, and deep-sky images. I also believe that pictures of observers and their equipment are essential in any book about observing the night sky. There are plenty such images here.

The biggest challenge for any writer in this field is to decide at what level to pitch the book. The other problem is what to leave out; publishing, like everything else, is governed by economic constraints. I hope I've got the balance right because, above all, this book is intended to inspire determined beginners and experienced observers alike. I have split the book into two parts, dealing with the basics and equipment first and then moving on to the observers and their techniques.

Complete beginners will want to read Chapters 1 to 3, but the experienced observer has my permission to start at Chapter 4!

Above all, I want this book to encourage and lead the potential new amateur astronomer on a voyage of discovery, ultimately emulating the remarkable people discussed in Part II.

Good luck on your voyage.

Martin Mobberley
Suffolk, United Kingdom
July 2004

Acknowledgments

I am indebted to many fellow astronomers who have freely donated images to this book. It is a policy I have always adopted myself; if anyone wants to use my images, they can have them – just drop me a line; I'm flattered that others want my images! To be honest, most amateur astronomy books would not be commercially viable if not for the generosity of amateurs in this way. But, even though I knew how generous my fellow astronomers were, I was staggered by the speed and enthusiasm with which they flooded me with their pictures!

I am indebted to the following astronomers:

Ron Arbour; Mark Armstrong; Tom Boles; Mike Brown; Denis Buczynski; Ron Dantowitz; Brad Ehrhorn; Andrew Elliott; Ray Emery; Nigel Evans; Steve Evans; Sheldon Faworski; Maurice Gavin; Ed Grafton; Karen Holland; Guy Hurst; Nick James; Chris Kitchin; Brian Knight; Weidong Li; Jan Manek; Pepe Manteca; Hazel McGee; Michael Oates; Arto Oksanen; Donald Parker; Damian Peach; Terry Platt; Gary Poyner; Tim Puckett; Gordon Rogers; Michael Schwartz; Greg Terrance; Roy Tucker; and James Young.

Many thanks also to the following companies and organizations:

Starlight Xpress; Meade Instruments Corporation; Celestron International; RC Optical Systems; Astrophysics; Scopetronix; IOTA-ES; SBIG; JPL/NASA; ESA; Apogee; Finger Lakes Instruments; and Lick Observatory.

I must also thank my parents for their support: in particular, my father's help in all my astronomical endeavors.

Finally, my thanks must also go to John Watson (London) and Louise Farkas (New York) and their respective teams at Springer, without whom this book would not exist at all!

Martin Mobberley
Suffolk, United Kingdom
July 2004

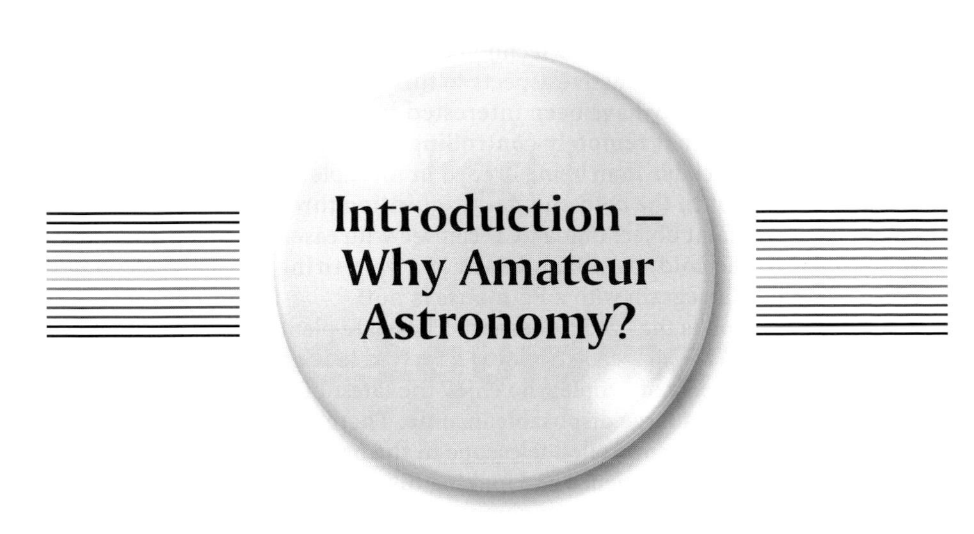

Introduction – Why Amateur Astronomy?

I have always felt that most people are fascinated by what is generally termed "Space." As an 11-year-old I, along with everyone else on the planet, watched in awe as Armstrong and Aldrin landed on the Moon. The human race has never been more transfixed. Who hasn't gazed up at the night sky and asked "Why are we here?" or "Is there anybody else out there?" Science fiction like *Star Trek* and *Star Wars* has always attracted millions of viewers and Sci-Fi comedy like Douglas Adams' *The Hitchhikers Guide to the Galaxy* has always been popular.

There is no shortage of TV programs about astronomy and space travel, either: satellite channels like *The Discovery Channel* cover the subject almost every day.

But, despite all this, amateur astronomy has generally been a minority pastime. It is not hard to see why: watching TV programs about space is one thing, but going out in the cold, dark, and damp is another. Training the eye and brain to recognize the subtle details on Jupiter, Mars, Saturn, or even Venus, is far from easy, and even "bright" galaxies and comets look like faint smudges to the beginner.

However, in the last ten years a new amateur astronomy has emerged, as if from nowhere. The old stereotype visual observer, peering through an eyepiece, has been joined by a newer and (sometimes) younger breed, who observe by watching a personal computer (PC) monitor inside a warm room, using powerful remote-controlled telescopes and CCD (charge-coupled device) cameras. And seasoned observers/photographers like myself have found our hobby enhanced beyond belief. Within minutes of capturing an image I can e-mail it to fellow observers worldwide and post it on my Web page; amateur astronomy has suddenly become a "cool" hobby! Many of my fellow amateurs simply take snapshot CCD images to show their friends, some strive to take the best images possible, in black and white and in color, while others carry out real science and make

CHAPTER ONE

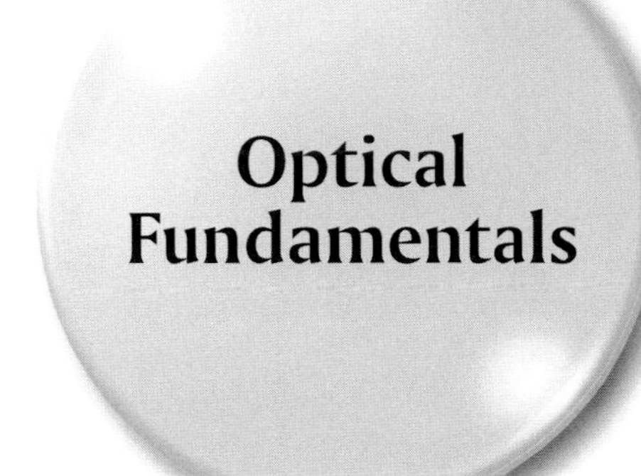

Optical Fundamentals

Visual Considerations

The purpose of any visual telescope is to collect and focus more light than the human eye, so that an eyepiece can be used to inspect and magnify the image formed at the focal plane. A large telescope will enable faint objects (stars, nebulae, galaxies, and comets) to be seen; it will also enable high magnifications to be employed on the Moon and planets, providing the Earth's atmosphere will allow it. In fact, magnification is an overused and abused term.

A telescope of focal length x, used with an eyepiece of focal length y, will have a magnification of x/y, but if this resulting value is more than several hundred, the chances are you will be looking at a magnified hopeless blur.

First, a telescope has to have a large aperture to have any hope of delivering sharp images at a high magnification. Second, even if the aperture is large, the atmosphere will often be so turbulent that there is little point using magnifications more than 100–200 times anyway. I know few amateurs who would ever use a magnification much greater than the telescope's aperture in millimeters.

The theoretical diffraction limited resolution (the "Rayleigh criterion") of any design of telescope, optical or radio, is determined by the formula:

$$\text{Resolution, in radians} = 1.22\lambda/D$$

where D is the aperture and λ the wavelength, measured in the same units.

There are 2π radians in 360 degrees, but many amateur astronomers will be unfamiliar with the concept of a radian, so let's change to the unit of the arc-second, which is 1/3600th of a degree, 1/60th of an arc-minute, or 1/206,265th of a radian. The formula now becomes:

$$\text{Resolution, in arc-seconds} = 251{,}643\lambda/D.$$

If we now set λ to the wavelength of green light, i.e., 550×10^{-6} mm, the formula becomes:

$$\text{Green-light resolution, in arc-seconds} = 138/D,$$

where D is aperture in mm. In other words, the resolution of a 138-mm aperture telescope will be one arc-second. This is very similar to the so-called Dawes limit for resolving double stars with an unobstructed refractor, namely:

$$\text{Resolution} = 116/D,$$

where D is, again, in mm.

So let's say we are using a telescope with an aperture in the 116–138 mm region, say, a 125-mm instrument. If atmospheric conditions are reasonable we would expect this telescope to resolve an arc-second, but what actual magnification should we use? A lot depends on the observer's eyesight and personal preference here. A good human eye can resolve one arc-minute, so, in theory, an eyepiece giving 60× magnification will just enable a one arc-minute eye to resolve one arc-second detail. However, in practice, twice this magnification would feel a lot more comfortable, i.e., 120×. With larger telescopes, the steadiness of the atmosphere will dominate the choice of maximum magnification.

Not surprisingly, a telescope will fail to deliver its theoretical diffraction-limited performance if the mirror or lens is not ground and polished to the correct shape. For a reflector, there should, ideally, be no deviation from the perfect curve, anywhere on the reflecting surfaces, by more than 1/8 of the wavelength of light; the corresponding accuracy for the lens surfaces of a refractor is only 1/2 the wavelength of light. In practice, a more meaningful measure is the RMS value, i.e., how smooth the whole surface is, on average, not just the worst deviations. The required RMS averages for perfection are nearer to 1/16 and 1/4 wave for mirror and lens surfaces, respectively. One-sixteenth of the wavelength of green light is approximately 35 millionths of a millimeter and yet, with care, telescope mirrors can be mass-produced at an affordable cost; I have always found this amazing!

If the surface accuracy of mirrors or lenses drops below the vital values, the contrast of features near the resolution limit will start to drop noticeably; in other words, close double stars will fail to be split and fine planetary details will not be seen at the highest magnifications. Not surprisingly, the major manufacturers pay very careful attention to optical quality: we are living in an age where astronomy magazines frequently report on how good commercial telescope optics are.

There is also a limit to the lowest useable magnification a telescope can support. The bundle of light emerging from a telescope eyepiece (the exit pupil) has a diameter which is equal to the telescope aperture divided by the magnification. Thus, a telescope of aperture 80 mm and a magnification of 10×

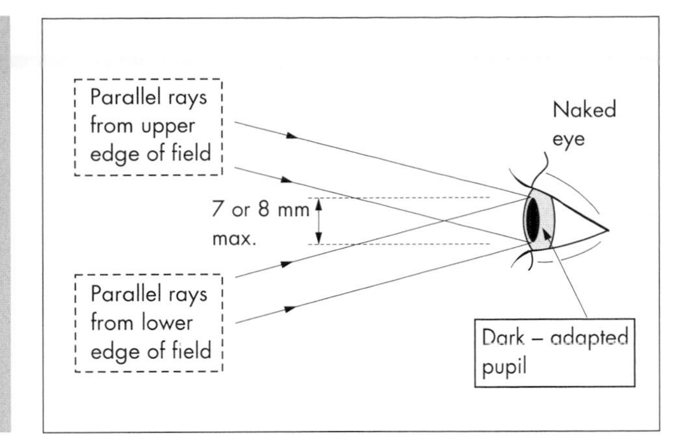

Figure 1.1. A telescope exit pupil larger than 7 or 8 mm cannot enter even the youngest eye.

will produce an exit pupil of 8-mm diameter; too large for even most young human eyes too accommodate, so light will be wasted (see Figure 1.1). In short, if you use a telescope *x* times bigger than the maximum aperture of the human eye, you must use an eyepiece which gives at least *x* times magnification, unless you do not mind wasting light (which may not be that important if you simply desire a very low magnification and a correspondingly wide field of view). In effect, the diameter of the dark-adapted pupil enforces a fundamental lowest useful magnification on a visual telescope. This prevents the observer from increasing the surface brightness of extended objects and seeing vibrant colors. However, there is no doubt that you will see far more detail in faint nebulous objects when using a large telescope; the increased number of photons and higher resolution will always pay dividends.

A sensible rule of thumb for the minimum useful magnification for a young person's telescope is $1\frac{1}{2}$ times the aperture in centimeters, e.g., 15× for a 10-cm telescope. A middle-aged person may well prefer 20× per 10 cm of aperture, as this corresponds to a 5-mm maximum dark-adapted pupil.

Of course, using a CCD camera, or even film, there is no pupil bottleneck and no exposure limit, so colors in faint objects can be recorded with ease.

In terms of how faint the observer can see (the limiting visual magnitude) with a given aperture, well, this could take up a whole book on its own. The standard formula is:

$$\text{Lim } m_v = 2 + 5 \log_{10} D$$

where *D* is, once again, the diameter of the telescope in millimeters. This formula predicts mag 12.0 for a 100-mm aperture, 14.0 for a 250-mm aperture, and 15.0 for a 400-mm aperture.

Speaking from my own experience, this formula is just right for telling me what is the faintest star I can see through a telescope. However, I know many experienced variable-star observers who can see a magnitude fainter, even in city skies! In the darkest skies, where background sky "noise" is not a problem, much fainter magnitudes have been claimed by experienced observers.

CCD Considerations

When a detector *other* than the human eye is employed, the purpose of the telescope is unchanged, but more powerful options emerge. In all cases, visual or CCD, the image size at the focal plane of a telescope is tan $\theta \times$ telescope focal length. The angle θ is the the angular size, e.g., the Moon is 0.5 degrees in diameter.

With a CCD, the focal plane image is sitting on the CCD surface, but with an eyepiece, the image at the focal plane is examined by the eyepiece, and the final image is focused on the observer's retina.

The human eye and brain is a remarkable combination, but transferring what is seen visually to a drawing is a big challenge and the retina does not have a long-exposure option! Nevertheless, an observer can walk out of a fully illuminated room into a dark backyard, and in seconds his or her eyes have become thousands of times more sensitive. The same observer can then switch from observing fine planetary details to observing faint deep-sky objects without any difficulty at all. Even CCDs have not yet become this versatile.

The eye has a huge brightness range over which it can operate and the brain is an excellent image processor, able to look at an object over many seconds and deduce what details are really there, despite the turbulence or atmospheric "seeing" and despite the object being very faint. In tests, I have found that a CCD exposure of a second or more is needed to capture all the faint nebulous detail seen by the eye, using the same telescope. As far as I am concerned (others may disagree) this is not because the retina has a multisecond exposure capability, it is because the eye–brain combination is excellent at extracting the signal from the noise over multiple "exposures." To the eye the noise is the background sky; to the CCD it is the background sky and the electronic noise caused by the CCD's temperature and the reading-out process. But CCDs do have properties which outperform the capabilities of the eye, the brain, or the human being. As already mentioned, CCDs can expose for minutes or even hours. The transfer of the data doesn't rely on the human's memory and drawing skills. Finally, and perhaps, most importantly, the CCD and computer don't get tired and cold. A PC and CCD can capture hundreds of galaxy images in a supernova patrol, or thousands of planetary images in the search for one sharp picture. The images captured can be stored for analysis the next day. Powerful image-processing tools are also available for extracting the most out of the images; the CCD, therefore, has many advantages. I will have much more to say about the specific advantages of CCDs in later chapters.

Telescope Configurations

There are three basic telescope configurations whose attributes are worth mentioning, although I don't intend going into great detail in a book primarily about new technology. I'm sure most readers of this book will be familiar enough with telescopes to skip the next section; please do, I won't be offended.

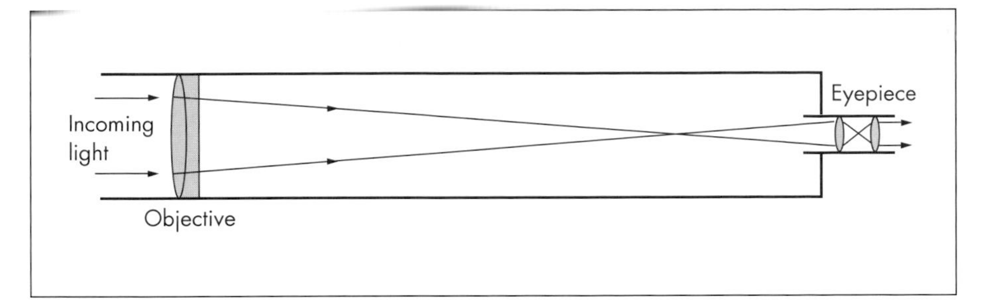

Figure 1.2. The refractor.

What I would stress, though, is that with all telescopes, the optics should be properly collimated and of a high quality, i.e., the mirrors and lenses should be lined up correctly; if they are not, don't expect good results!

Refractors

Firstly, let us look at the refractor (Figure 1.2). The quality refractor uses an achromatic lens doublet to focus the incoming light and minimize chromatic aberration (i.e., the splitting of light into a spectrum). Refractors have fixed collimation (i.e., the optics can't move about) and closed tubes (so air currents inside the tubes are almost nonexistent). In small apertures they are rugged and transportable. However, they are very expensive for their aperture and, unless expensive apochromat designs are used, they cause colored fringes around bright stars, planets, and features on the Moon. However, small-aperture apochromats are highly prized telescopes, especially when the observer requires a wide field of view or an instrument that can be carried abroad for, e.g., an eclipse expedition.

Newtonian Reflectors

Newtonian reflectors use a mirror to focus the incoming light (see Figure 1.3a). They suffer from tube currents, they can easily go out of collimation, and they have a second mirror at the top of the tube to divert light to the eyepiece. The presence of this secondary mirror can slightly degrade the contrast if the secondary is large, but this is only a real concern for planetary observers. The reflector is excellent aperture value for money, especially in the alt-azimuth mounted Dobsonian format discussed shortly. Reflectors suffer from an aberration called coma, whereby stars start resembling seagulls at the edge of the field. This aberration becomes worse with decreasing f-ratio. Despite all these disadvantages, a long-focus (f/6 or higher) Newtonian, with well-collimated optics and a well-ventilated tube, is a joy to own and can deliver superb results. Short-focus Schmidt–Newtonians, which feature a corrector plate to reduce coma (see Figure 1.3b), have

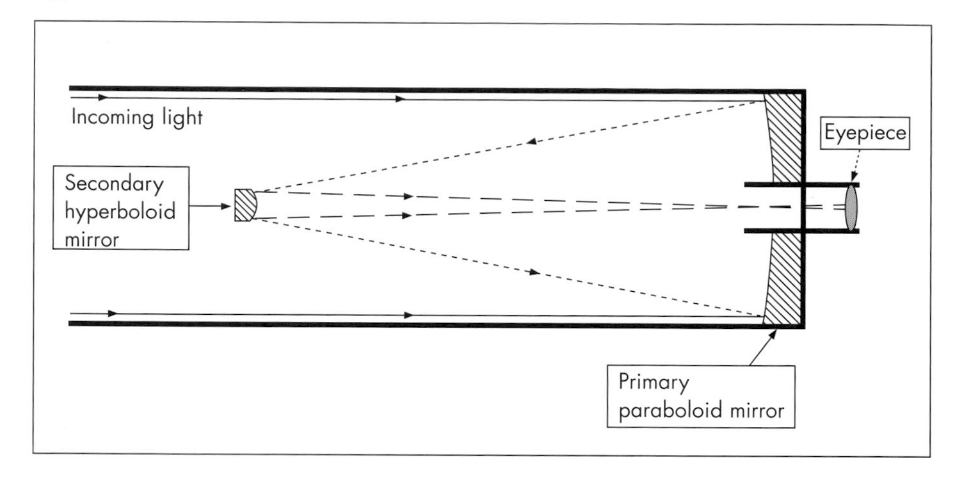

Figure 1.5. The Cassegrain reflector.

amplifies this to f/10. Aberrations inherent in such a fast f/2 design are reduced by the use of a two-sided aspheric Schmidt corrector plate at the end of the tube (Figure 1.6). This plate has the additional advantage of sealing the tube, thus minimizing tube currents and acting as a platform on which the secondary mirror can be mounted. The primary mirror of a commercial Schmidt–Cassegrain is, typically, 3 or 4 percent larger than the telescope aperture; this guarantees that the whole of the field of view is fully illuminated. The secondary mirror is of a

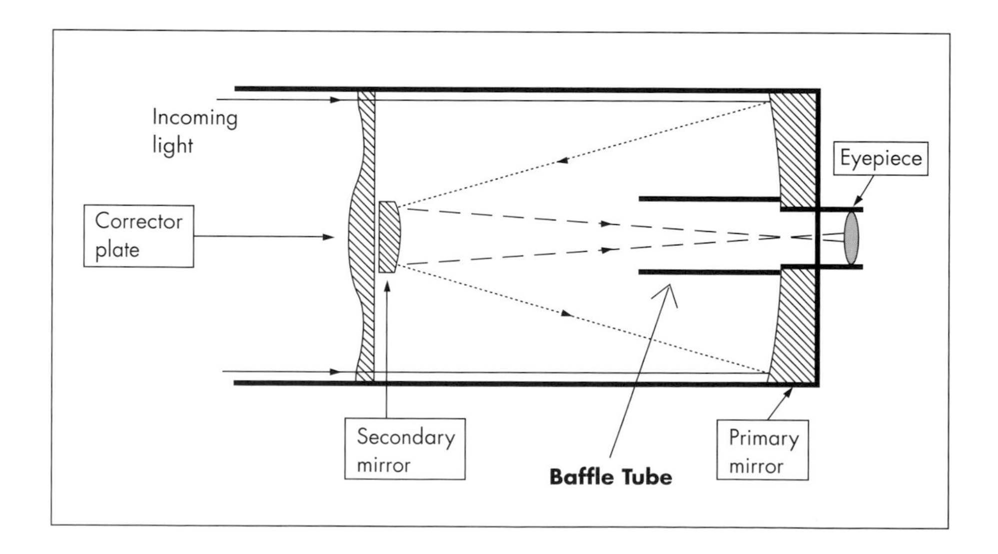

Figure 1.6. The Schmidt–Cassegrain.

convex, aspheric design. For optimum performance the optical system of a Schmidt–Cassegrain must be well designed.

In any Cassegrain system, there is a risk of stray light from the sky sneaking past the secondary mirror support and flooding the field of view; the risk is especially high when the Moon (or a total solar eclipse at the diamond-ring stage) is being observed. Making the secondary mirror support very large will prevent this but will cut off more light than is desirable; it will also result in unpleasant diffraction effects, thereby degrading optical performance. In addition, the design of the primary and secondary baffle tubes is also critical. Both Meade and Celestron use strategically placed field stops in their primary baffle tube to prevent scattered light degrading the image. If you were to have a Schmidt–Cassegrain made as a custom "one-off" instrument it would cost at least five times the current commercial price (depending on whether you imported the instrument or not).

Mass production for the American market has brought this compact design within the reach of many amateurs. The two major players in the Schmidt–Cassegrain market are Celestron (since 1970) and Meade (since 1980). My own 30-cm Meade LX200 is shown in Figure 1.7.

Originally, in the 1970s and up to the late 1980s, Celestron were the market leaders and pioneered the Schmidt–Cassegrain revolution. In the 1960s Celestron's Thomas J. Johnson found a way of mass-producing the complex optics and corrector plate. Here was an ultra-compact telescope with a tube length barely twice the mirror diameter and a range of accessories to enable long-exposure deep-sky astrophotography. Now, in the twenty-first century, the range of accessories is staggering.

Figure 1.7. The author and his 30-cm Meade LX200.

Modern eyepieces, from companies like Celestron, Meade, and TeleVue, are of a very high quality and, in general, you pay more for a larger field of view. Let's just be clear about one thing though: the real field of view equals the apparent field of view divided by the magnification. Thus, if you have an eyepiece with a 50-degree apparent field presented to your eye, a magnification of 100× will give you a real field (on the sky) of half a degree.

For the widest fields we need to consider whether to use a standard 1.25-inch (31.7-mm barrel diameter) eyepiece or a 2-inch (50.8-mm barrel diameter) eyepiece. The advantage of a 2-inch eyepiece is that the internal barrel diameter is approximately 46 mm in diameter, as opposed to 27 mm in the 1.25-inch diameter eyepiece.

Eyepieces with longer focal lengths (i.e., lower magnifications) and wide apparent fields will have their fields compromised by the smaller barrel diameter.

We can make the crude assumption that the eyepiece barrel merely needs to accommodate the diameter of the focal plane image, so that a handy formula can be derived which will tell us what size of eyepiece barrel we need: If we call the diameter of the image formed at the telescope focal plane, D (in millimeters), then

$$D = F_A \times \mathrm{FL.}/57.3$$

where F_A is the apparent field of the eyepiece in degrees and FL is the focal length of the eyepiece in millimeters.

Turning this formula around and setting D to 27 mm (the maximum internal diameter of a 1.25-inch eyepiece) gives us the maximum focal length (MFL) eyepiece we can use with the smaller eyepiece size, i.e., $\mathrm{MFL} = 27 \times 57.3/F_A$.

Thus for 31.7-mm barrel eyepieces a 50-degree apparent field design implies a MFL of 31 mm; a 65-degree apparent field design implies a MFL of 24 mm; and an 84-degree apparent field design a MFL of 18 mm. A brief glance at the available eyepieces on the market confirms this. For example, Tele Vue's 68-degree field Panoptic range eyepieces switch to a 2-inch barrel for focal lengths above 27 mm and their 82-degree Nagler range eyepieces switch above 16 mm.

As the eyepiece apparent field and focal length set the barrel size requirement the amateur does not need to take the telescope into consideration.

In practice the maximum real fields available with today's eyepieces (i.e., filling a 2-inch barrel field lens) are with 55-mm, 50-degree field Plossls or 40-mm, 67-degree Super Wide Angle eyepieces. If you have deep pockets, the 31-mm, 82-degree TeleVue Nagler is a maximum real and apparent field eyepiece!

But before rushing out to buy a 55-mm, 50-degree eyepiece, remember our earlier look at exit pupils. Unless your telescope has an f-ratio of f/7 or more the exit pupil will exceed 8 mm, a bit too much for even the youngest eye! A 2-inch barrel, 55-mm, 50-degree eyepiece with, say, a 180-mm, f/9 refractor, would be mouth-watering (29× mag, 6-mm exit pupil, 1.7-degree real field, and pin-sharp stars even at the field edge at f/9).

To calculate the maximum possible real field ($x°$) on the sky for a given telescope focal length, the following formulae are sufficiently accurate for the small angles involved.

For 1.25-inch (31.7-mm) barrel eyepieces:

$$x° = 27 \times 57.3/\text{Focal length in mm} = 1547/\text{Telescope focal length in mm.}$$

For 2-inch eyepieces:

$$x^\circ = 46 \times 57.3/\text{Focal length in mm} = 2636/\text{Telescope focal length in mm.}$$

Magazines like *Sky and Telescope* and *Astronomy* always contain volumes of eyepiece adverts. Every eyepiece I have used by Celestron, Meade, or TeleVue (the latter being the very best) has delivered a quality image, but you must expect to pay at least $50 for even a basic 50-degree apparent field Plossl; any less and you simply won't get a good eyepiece. Most keen amateur's favorite eyepieces will have cost them at least $150 and probably more.

throughout the world is truly amazing, but so is the number appearing in second-hand ads! Perhaps this is because the percentage of buyers using their telescopes to their full capacity is tiny. Maybe something is wrong somewhere.

In my experience, the biggest single reason for amateurs not using their equipment is simply a lack of spare time. The unpredictability of cloud is another big factor; you can't say "I'll reserve Wednesday night for observing"; it just doesn't work like that. The third factor, and something you can do something about, is having equipment that is always ready to roll, easy to use, and just outside the back door. I can't emphasize this last point enough. Any telescope that is intended for regular use should if at all possible be permanently mounted in the owner's backyard in an observatory that is simplicity itself to open up. If you can arrange this, then within ten minutes of leaving the house, observations can be made and images collected. The observatory pictured in Figure 2.1 was specifically designed for ease of use and has been fully described in *More Small Astronomical Observatories*, in this series. It is edited by Patrick Moore and was published in 2002.

Figure 2.1. The author's user-friendly LX200 run-off shed.

My observatory is a lightweight, compact run-off shed which glides freely up and down on its rails, is a joy to use, and has transformed the ease with which I carry out astronomy. Obviously, the compact dimensions of the 30-cm Meade LX200 inside the shed played a major part in minimizing the size of the shed.

Recommendations

This conveniently leads me to my definite recommendation about buying a commercial telescope. If you have astrophotography/imaging aspirations and $2000 to spend, consider buying a modern Schmidt–Cassegrain *from a reputable dealer*.

I have used my LX200 far more than the 36- and 49-cm Newtonians I also own; it is simply far easier to use and the Go To facility, which slews the scope within 5 arc-minutes of the target, saves so much time.

Years ago, Schmidt–Cassegrains had a poor reputation for high-resolution work, but now most of the world's leading planetary imagers and supernova patrollers use them. With modern CCDs and accessories they are unbeatable technological value for money. Perhaps their biggest advantage is their compact size; with a 30-cm Newtonian, the eyepiece can end up almost anywhere, a set of steps or even a stepladder often being required, whereas a Schmidt–Cassegrain eyepiece is nearly always conveniently placed.

I will stress again, **a telescope that is easy to use is the best telescope to own**, and Schmidt–Cassegrains are the easiest medium-aperture telescopes to use. The leading planetary observer Damian Peach, who has never owned an observatory, is shown with his 28-cm Celestron Schmidt–Cassegrain in Figure 2.2.

Okay, so that's my recommendation for amateurs with $2000 and more to spend, but let's come down to earth for a moment and consider a newcomer with a more limited budget.

If you are spending between $200 and $1000 and are primarily interested in visual observing, take a serious look at a Dobsonian reflector. Dobsonians are simply Newtonian reflectors mounted on a low-friction alt-azimuth mounting. They cannot be used for long-exposure photography unless heavily modified. They are named after the Californian ex-monk John Dobson, who has spent his life promoting the construction and use of these "easy-to-use" reflectors. As well as being user-friendly, they are an extraordinary value for the money.

In 2003, $200 will get you a 114-mm (4.5-inch) Dobsonian; $350 will get you a 150-mm Dobsonian; and $480 will get you a 200-mm (8-inch) Dobsonian. Larger apertures are also readily available and a 250-mm (10-inch) Dobsonian, often considered the best portability/aperture compromise, will only set you back $650. Many experienced amateurs will tell you there is no substitute for aperture, and Dobsonians provide the biggest aperture for your money.

That's largely because Dobsonians also have no motors or electronics. There's almost nothing to go wrong. It is possible to get digital "setting circles" to fit to the altitude and azimuth bearings of a Dobsonian to tell you where to push it to find the object you are looking for. Gary Poyner, one of the world's leading variable-star observers, uses a large 45-cm Dobsonian to make up to 12,000 variable-star estimates per year! A modern Meade Dobsonian is shown in Figure 2.3.

Figure 2.2. Planetary imaging expert Damian Peach and his 11-inch (28-cm) Celestron Schmidt–Cassegrain at his former UK home in Rochester, Kent. Note the lack of any observatory! Photo: courtesy Damian Peach.

I've described the various Schmidt–Cassegrain models in Chapter 4 and also the amazingly portable ETX Maksutovs from Meade. Meade's 90-mm ETX has been a runaway commercial success because it combines total portability with a $500 dollar price tag; however, a 90-mm aperture would leave many amateurs feeling starved of light!

You may also be tempted to buy an equatorially mounted reflector from a major manufacturer. Up to the 1980s, a large-aperture equatorially mounted Newtonian was the amateur's dream telescope. But that was before Schmidt–Cassegrains became so competitively priced and mass-produced. It was also in an era when amateurs wanted a wide visual field of a degree or more for when they were star-hopping to a target. Plus, in those far-off days, many amateurs built their own telescopes, *including* the optics!

In the twenty-first century, commercial Newtonians appear to have split into two categories, namely: the alt-azimuth mounted Dobsonians and the equatorially mounted short-focus Schmidt–Newtonians.

One thing Schmidt–Cassegrains *can't* deliver is a wide field of view. Many visual observers crave a 2-degree field, and can get this from a short-focus Newtonian. And a system with a focal length of less than one meter will give a

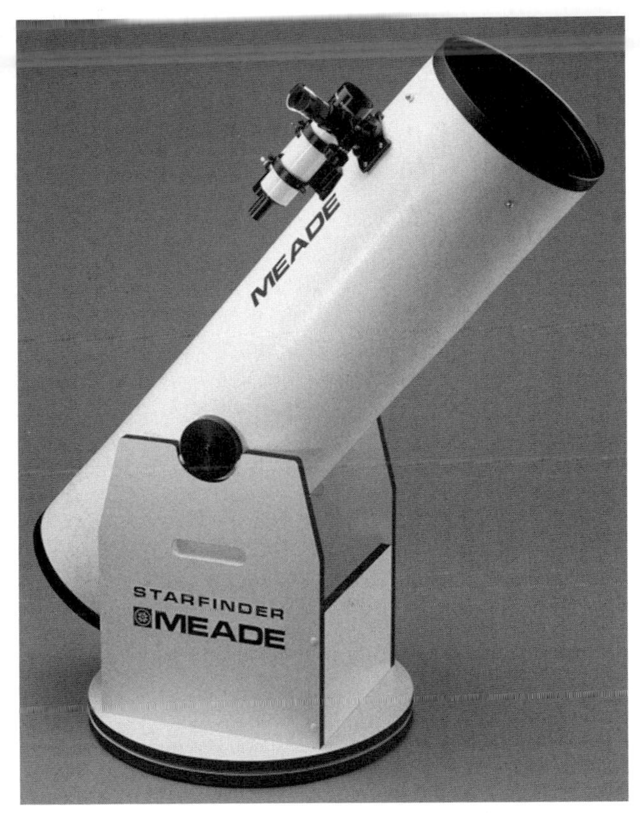

Figure 2.3. A Meade 12.5-inch (32-cm) Dobsonian. Photo: courtesy Meade Instruments Corporation.

CCD field of view of some 20 arc-minutes, even with the smallest CCD detectors. Wonderful for when those bright comets come along!

That's the philosophy behind Meade's excellent LXD55 series Schmidt–Newtonians.

The "ultimate Newtonian" is probably the long-focus planetary instrument; I'm talking here about f/7–f/10 Newtonians, of 25–40 cm aperture. Mirrors of these very long focal ratios can be ground and polished to perfection and are far easier to collimate than short-focus reflectors. They can also incorporate the tiniest secondary mirrors, to minimize diffraction effects. The views they can provide of the planets are exquisite, but no-one makes them commercially, because they are just too long to mount solidly at a reasonable cost.

I think I've said enough about buying a commercial telescope, but I would just like to offer a few bullet points:

- Buy a telescope from a national main dealer.
- Read the equipment reviews in the astronomy magazines like *Sky and Telescope*, *Astronomy Now*, or *Astronomy*.
- Search the Internet for telescope discussion groups.
- Visit your local astronomical club or society and ask to look through members' telescopes before you make a decision.

- Accept that you will always regret not having bought a bigger aperture!
- Mount your new telescope permanently if at all possible, in an easy-to-use observatory with as few bushes and trees obstructing the view as possible.
- If you have less than $200 to spend, keep saving.
- If you are a visual observer with less than $2000 to spend, think Dobsonian.
- If you have CCD imaging aspirations, think Schmidt–Cassegrain.

There is, of course, one option I haven't touched on – buying a second-hand telescope. To be honest, it's a lottery. One reason for selling a telescope is that it is a pile of junk, even if it still looks glossy and new. Equally, another reason is just that the owner doesn't have the time to use it.

Personally, I would always insist on a demo of (or opportunity to try) any second-hand telescope before making a purchase. You'd test-drive a second-hand car, so why not a telescope? If you can take along a member of a local astronomical society or knowledgeable friend, even better. There are certainly second-hand bargains to be had, but care is needed – remember, "buyer beware"!

CHAPTER THREE

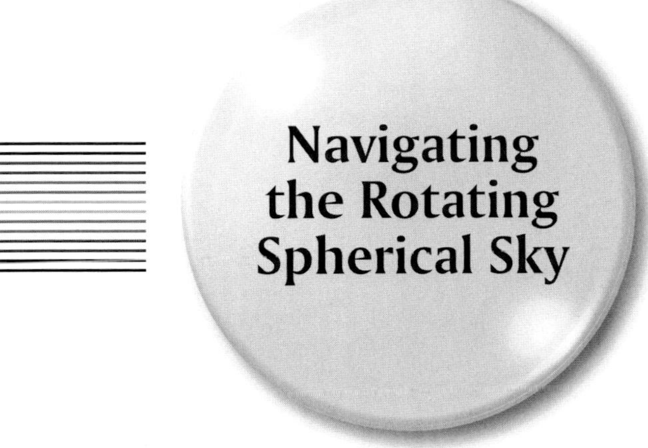

Navigating the Rotating Spherical Sky

The new amateur astronomer doesn't need to use detailed star charts and "star-hop" his or her way around the sky any more. The ability to recognize a dozen stars and the funds to purchase a modern "Go To" telescope is probably enough to get you to any object above the horizon (trees permitting). However, a basic understanding of just what we are seeing when we walk outside is of fundamental importance, whether you're using a high-tech scope or a Dobsonian. So before we come to the high-tech toys of the next chapter, let's have a quick grounding in what the night sky is all about.

With the exception of aeroplanes, meteors, and artificial satellites, everything in the night sky is a very long way away and everything rises in the east and sets in the west. The Earth rotates from west to east in 24 hours and so the stars and planets appear to go the other way in 24 hours (actually, 23 hours, 56 minutes, and 4 seconds, with respect to the stars, but let's not split hairs at this point).

The Moon goes around the Earth (new moon to new moon) every 29.5 days, so it rises an average of 49 minutes later each night and drifts slowly eastward amongst the stars. Everything else that's bright (except for a few comets) appears fixed to the dome of the sky. The easiest way for the beginner to visualize this is to take a globe of the Earth and imagine that the Earth is at the center of a much larger globe on which the stars are painted on the inside. This larger globe rotates in the opposite direction to the Earth and so allows stars to rise in the east and set in the west. If the Earth beneath you was transparent you could see the whole sky (except that the Sun would be permanently visible too), but if you could turn the Sun and Moon off and sit on an Earth made of glass, you could see all the stars in the sky in one go.

Of course, on a dark, moonless night you really can see 50 percent of the sky at any one time, but depending where you are on the Earth your view of the celestial

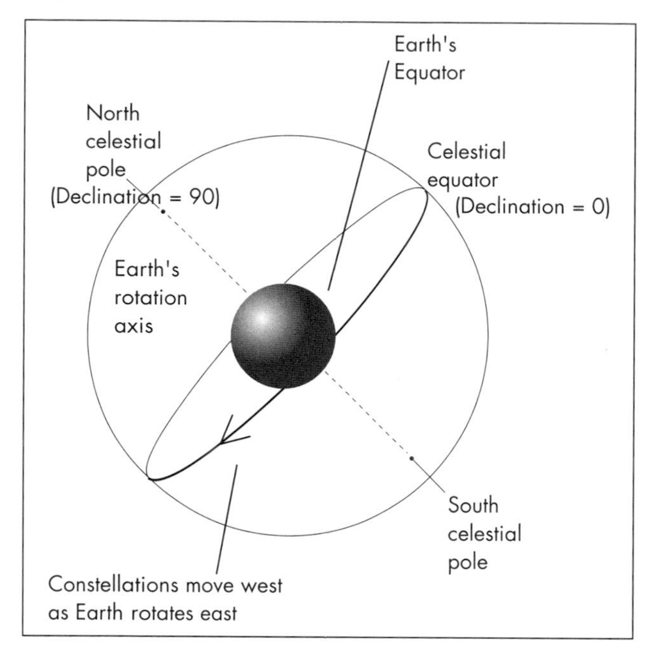

Figure 3.1. The Celestial Sphere: an imaginary sphere enclosing the Earth.

sphere will always be restricted in some way by the body of our planet. The celestial sphere is illustrated in Figure 3.1.

An observer at the north or south poles of the Earth will always see the same 50 percent of the sky (when the Sun is well below the horizon that is). At these extreme latitudes the stars don't rise or set, they just go around the horizon, always at the same altitude. At the northern pole the north-pole star Polaris will always be dead overhead.

An astronomer at the equator has the best overall view. The north-pole star Polaris sits on the north horizon, the southern polar constellations sit on the south horizon, and during a year the observer can view the whole sky. Obviously, any telescope that has to track the stars must be driven westward at a rate of one rotation every 23 hours 56 minutes and 4 seconds. This equates to a degree in every 4 minutes of time, or 15 arc-seconds per second of time.

In practice, the movement has to be smooth and accurate. If the motors, gears, and software that achieve this are not of a high enough quality, stars will trail badly on long exposures. An alt-azimuth mounted telescope like a Dobsonian can't track the stars (it just has altitude and azimuth axes, allowing it to move up and down, and left and right). Although this may seem a big disadvantage, a good Dobsonian with smooth bearings is a joy to use as only the merest touch will recenter the field. Such telescopes are superb for low-power deep-sky work.

An equatorially mounted telescope has its "azimuth axis" tilted over until it points straight at the north or south pole of the sky (depending which hemisphere you live in).

For an observer living at the north or south pole, a Dobsonian is already equatorially mounted! For every other latitude on Earth, the angle that the "azimuth axis" – correctly called the *polar axis* when aligned in this way –

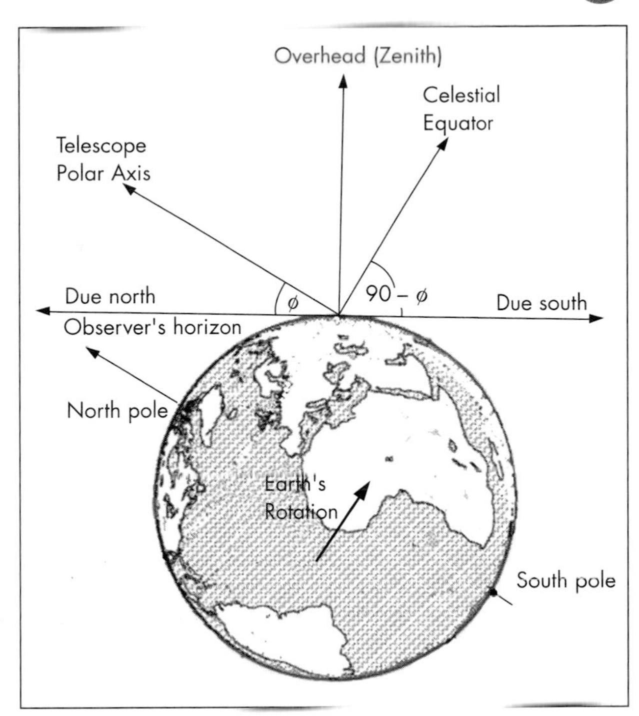

Figure 3.2. The view of the sky from a specific observer's latitude.

makes with the ground must be equal to the observer's latitude. In other words, at the Earth's poles the polar axis tilt is 90 degrees, at the equator it's 0 degrees, in New York it's 41 degrees, and in London it's 51 degrees.

The diagram in Figure 3.2 makes the far northern latitude observer's situation clear.

On the celestial sphere the equivalents of latitude and longitude are respectively called declination (abbreviated to Dec) and right ascension (abbreviated to RA). The right ascension concept sometimes puzzles newcomers. The declination of an object is measured in degrees, minutes, and seconds of arc, just as with latitude.

The odd name comes from early telescopes, which were almost all built in the northern hemisphere. If you face north while standing next to such a telescope, anything on the *right*-hand side of the polar axis rises, or *ascend*s, while the telescope is tracking the stars. Thus, right ascension (RA).

Right ascension is measured in units of sidereal (star) time, i.e., hours, minutes, and seconds. Again, that's because early telescopes used a clock mechanism to turn the polar axis. A complete east–west journey through the sky encompasses 24 hours. If you look at your local meridian (the north–south line which passes overhead) and find that a star with an RA of, say, exactly 7 hours is there, then after one hour an object of 8 hours RA will be at the same spot. Right ascension is just like the hour hand of a clock.

Well, almost! It's time to explain that little anomaly of the fact that 24 hours of RA goes by in 23 hours, 56 minutes, and 4 seconds. That is because the Earth is not only rotating, it orbits the Sun in the course of a year.

As the winter constellations are replaced by those of spring, then summer, then autumn, a complete cycle takes place in 365.256 days, a cycle equivalent to a shift of 24 hours, or 1440 minutes in time. Dividing 1440 by 365.256 days gives us 3.94 minutes, the amount that the sidereal time clock gains on real time every day.

In generations gone by, every proper observatory had a sidereal clock installed which indicated the right ascension of objects which were transiting the meridian at that time, that is, the "local sidereal time." (Most computer-controlled telescopes still will show local sidereal time on their hand-controller displays. They calculate this from the date, time, and longitude that you entered when you first initialized the telescope.)

So (I hear you ask!), if sidereal time/RA runs from 0 hours to 23 hours and 59 minutes, where is the starting point, where is 0 hours located and how can I use RA and Dec to find my way round the sky?

The 0 hour point for RA on the celestial equator (at zero degrees declination) is located at the vernal equinox, which is the point where the Sun appears to be when it moves from the southern to the northern hemisphere, in the northern hemisphere spring (around March 21). This used to be called the "first point of Aries," though it's now located in the constellation of Pisces! Not surprisingly, with the Sun on top of it, this point is totally unobservable in March! However, six months later, the 0 hour RA line transits the meridian at midnight (around September 21) and the constellations either side of it (Andromeda, Pegasus, Pisces, Aquarius, etc.) are all well placed. It follows from this that, three months later, the constellations straddling the six-hours line are transiting at local midnight, e.g., Orion in mid-December; in mid-March, 12 hours RA and the Virgo region is doing the same; ditto for 18 hours RA and Hercules/Ophiuchus in mid-June.

In practice, whether you are using mechanical setting circles or a Go To telescope, you will find your way around by calibrating the setting circles or telescope's brain on a known star while the drive is running and then slewing to the desired object. For example, on the original Meade LX200, the bright star Vega was designated the reference "Star 214.' By centering this star in the eyepiece and pressing "Star', "214', "Enter" and then holding the "Enter" key down till the keypad beeped, the LX200 was informed that it was on the star Vega. It then knew that it was pointed at RA 18 hours 37 minutes and Declination +38 degrees 47 minutes. (Note that RA minutes (of time) are much larger distances than Dec minutes (of arc) – 15 times larger at the celestial equator.) From this point, the telescope knew where it was and, providing its polar axis was aligned accurately and the telescope drive was working, it could then slew to any other part of the sky with an accuracy of ±5 arc-minutes or so.

Finding the RA and Dec of any star is easy. Any modern star atlas will contain a grid of RA and Dec lines from which an approximate position can be derived. Accompanying tables will then give you the precise RA and Dec for the brighter stars.

Planetarium software like Software Bisque's *The Sky*, or Project Pluto's *Guide 8.0* will enable you to get the RA and Dec simply by clicking the mouse on an object. What could be easier?

Jargon can be a puzzle to newcomers. For example, what are the *circumpolar* stars? Easy: these are stars which, by virtue of the observer's latitude, never set. They are always above the horizon and always visible on a clear night, circling the celestial pole.

For an observer at a latitude of x degrees, any star with a declination greater than $90 - x$ degrees will never set. Thus, for a location at 50 degrees north, stars further north than 40 degrees Dec will never set; they will circle the pole star, as shown in Figure 3.3a. A long exposure of the northern polar region, taken by Nigel Evans, is shown in Figure 3.3b.

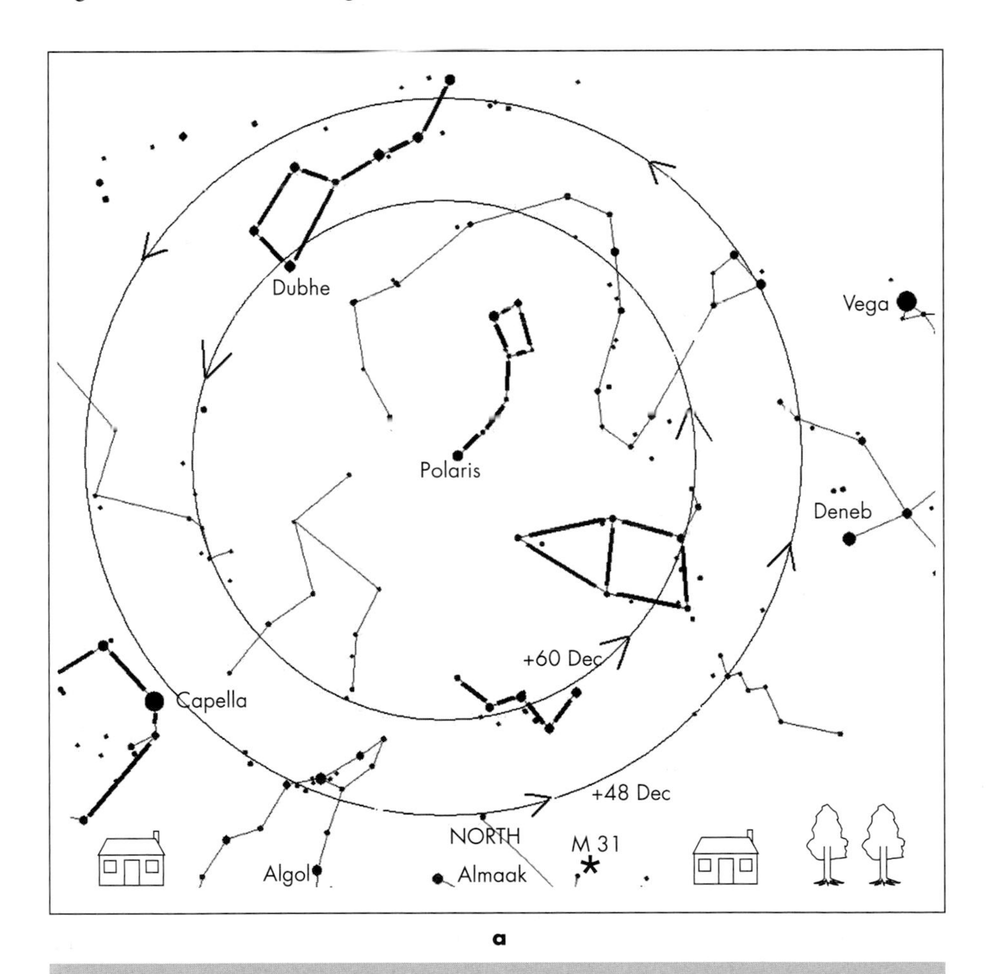

a

Figure 3.3. a Circumpolar stars are permanently above the horizon. For an observer at latitude 50 north, any stars at higher declinations than 40 north (90 – 50) will always be above the horizon, orbiting the north pole. The stars Algol and Almaak will just scrape the north horizon at their lowest point.

(Figure 3.3. b, see overleaf)

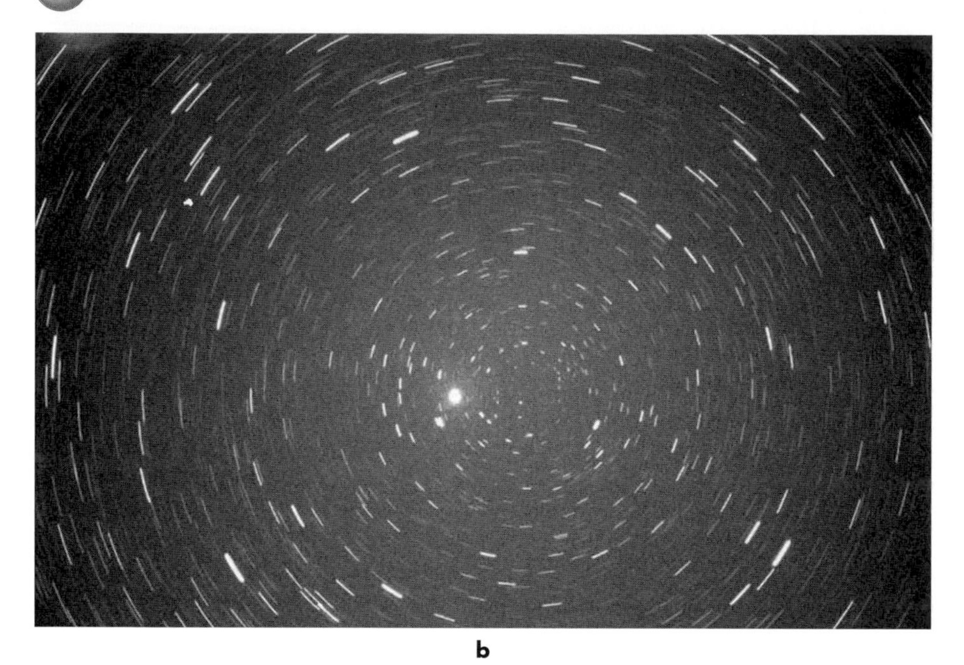

b

Figure 3.3. b Stars trail around the north celestial pole in a 20-minute exposure. Photo: courtesy Nigel Evans.

For a southern hemisphere observer in Sydney, Australia, at 34 degrees south, stars further south than −56 degrees Dec will never set; they will circle the south celestial pole.

The corresponding question is, which stars will never rise? For an observer at a latitude of x degrees, any star with a declination less than $x - 90$ degrees will never rise. Again, for a latitude of 50 degrees north, stars of declination −40 will just meet the south horizon when their RA is on the meridian, assuming a perfect transparency and no trees or buildings in the way (Figure 3.4). Anything further south will *never* be seen from this location. For our Sydney observer at 34 degrees south, stars of declination +56 will just meet his or her north horizon when their RA is on the meridian. Anything further north will never be seen. In practice, stars this low down will probably be invisible anyway because of atmospheric extinction.

Up to this point I have (deliberately) given the impression that only equatorially mounted computerized telescopes can slew around the sky and track objects. This is not the case (sorry, I thought it best to start simple). In fact, alt-azimuth mounted but computer-controlled "Go To" telescopes convert RA and Dec to altitude and azimuth. Instead of following an object by driving a polar axis westwards, they drive the telescope in *two* axes to follow the azimuth and altitude of a star. They cannot easily be used for long-exposure imaging – although there are

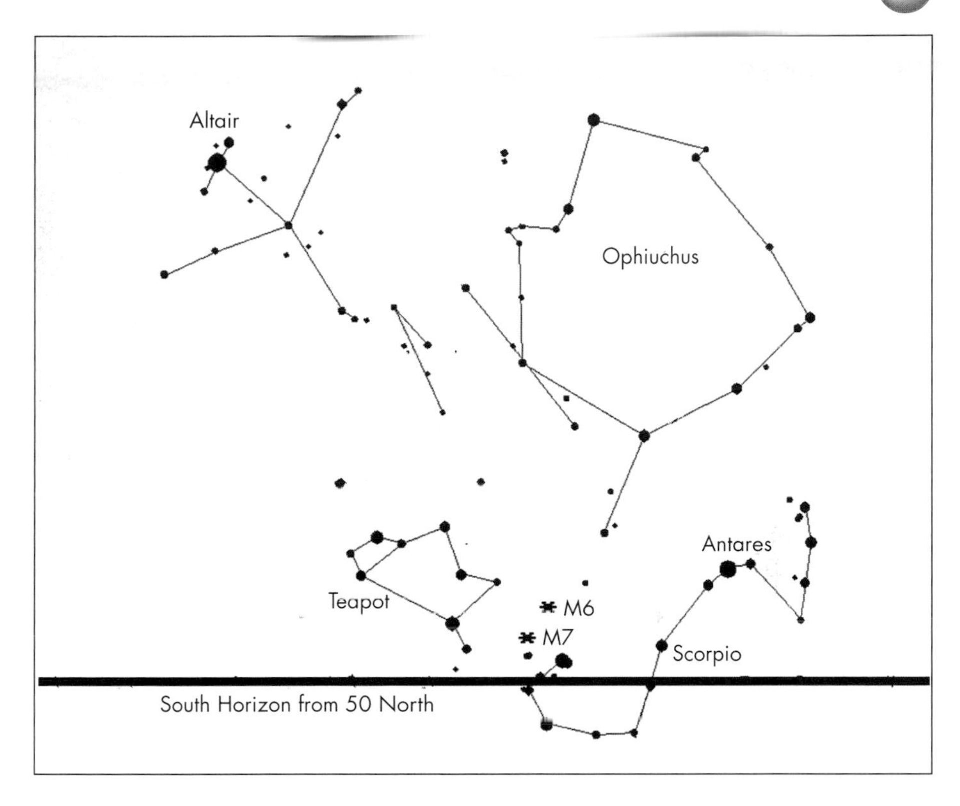

Figure 3.4. Stars of the southern hemisphere, just rising, above a northern hemisphere observer's south horizon. From 50 degrees north, stars of 40 degrees south will, in theory, just scrape the observer's southern horizon. Bright stars of 39 degrees south will (in theory) rise briefly above the horizon, but stars of 41 degrees south will never be seen. In practice, the bright star Antares, in Scorpio, at Dec –26, will be one of the furthest south stars that an observer at 50 north will ever see.

solutions even to this – because the field of view in the eyepiece slowly rotates (no, *you* figure it out!) but for visual use they are very convenient.

It's time to examine Go To telescopes in detail.

CHAPTER FOUR

"Go To" Telescopes and Mountings

Background

Since the mid-1980s, amateur astronomers have been able to buy commercial telescopes that will slew automatically to a desired target, eliminating the hassle of having to "star-hop" to the target or use setting circles in the dark.

In 1986, Celestron, the pioneers of the commercial Schmidt–Cassegrain, brought out their Compustar range. This system coupled a microprocessor to a stepper motor drive enabling the telescope to store the positions of 6000 objects and drive the motors to the desired target. However, this feature alone did not tempt many amateurs to part with the cash to buy a Compustar. The computer-controlled telescopes were much more expensive than their "dumb" counterparts, and affordable CCD cameras were some years away.

This latter point was, in retrospect, critical, as it would become evident that a new breed of high-tech amateur astronomers was waiting in the wings. Many of these amateurs would aspire to observing from indoors, and had little interest in braving the cold, dark, and damp. While a remote-controlled telescope was – in theory – available in 1986, it was not much of an attraction to visual observers and, without CCD equipment, of little interest to the high-tech generation.

These days there are plenty of computer-controlled commercial telescopes out there and their prices and specifications are constantly changing, even as this book goes to press. At the very least this chapter gives a reasonable overview of what the big manufacturers of computerized telescopes were offering in 2003.

The LX200 Changes the World

In 1992, the Meade Instruments Corporation unveiled their answer to the Compustar, the soon-to-be-ubiquitous LX200 Schmidt–Cassegrain. By the early 1990s CCD cameras were on sale and, within a few years, accessories would become available to make remote observing relatively hassle-free: accessories such as telecompressors which would shorten the focal length of the telescope (until the CCD chip field-of-view was much larger than the slewing accuracy), autoguiding CCD chips which could ensure near-perfect tracking of the telescope, and planetarium software (just click on a star map and the telescope just points there!). But the real strength of the LX200 was its competitive price; by using inexpensive drive components and optical encoders (instead of the stepper motors of the Compustar) the LX200 was affordable. Periodic error-correction software corrected the inevitable inaccuracy of the drive train, so that the system performed as well as much more expensive systems.

The rest is history: the LX200 became the best-selling serious amateur telescope in the world.

The Meade ETX

In 1996 Meade developed another best-seller, and its compact ETX became a runaway success story. At $500 for a 90-mm Maksutov with sidereal drive and good optics, what else could it be? But Meade decided to go one step further by introducing a Go To facility on this tiny portable telescope, for only $150 more!

Unlike the LX200, this little telescope was not designed for advanced amateurs with CCD imaging aspirations; it was intended more for visual observers who wanted to “sight-see” around the sky with a few button-presses, and not break the bank doing it.

The 90-mm ETX Autostar (the Go To version of the ETX) has also been a bestseller.

I have owned a Meade ETX for many years and am constantly amazed at the optical quality and value-for-money of the instrument. However, to get the ETX to slew accurately I would advise a systematic approach. There is no way that a small-diameter worm wheel can *guarantee* to slew a telescope to a target within plus or minus half a degree (however clever the software), but the ETX has a finder, and will drop the desired object into the finder’s field, and usually in the main field as well.

Aligning the telescope with north and levelling the base (in alt-azimuth mode) or aligning with the pole (in polar mode) needs to be done carefully and the friction clutches must be well tightened to avoid slippage. If you accept that your targeted object won’t necessarily be visible in the ETX eyepiece, but will at least be in the finder’s field, you will find this instrument comfortable and enjoyable to use.

Figure 4.1. a The Meade ETX 105 EC with standard controller. Photo: courtesy Meade Corporation.
(Figure 4.1. b, see overleaf)

a

The ETX Maksutov also comes in 105- and 127-mm apertures (see Figure 4.1a) and with the latter it comes head-to-head with Celestron's NexStar 5i, which I will talk about soon. There are also 60- and 70-mm short-focus ETX refractors (see Figure 4.1b).

As an owner of a 127-mm Meade ETX – along with its tripod it weighs only 32 pounds (14 kg) – I find I can lift the whole assembly outside to observe in one go. I keep the system fully assembled by the outside door, so in less than five minutes I can be outside and observing. A 20-cm SCT (for example) on an equatorial mounting would just be too heavy to do this, and anything much smaller than a 127-mm aperture would not give enough aperture, for my needs, at least.

I find the system shown in Figure 4.2 is pretty well optimum for casual, convenient observing. Although I own much larger telescopes, I use my 127-mm ETX a lot because it is so convenient. I wonder how many advanced amateurs actually spend time enjoying the simple pleasures of the night sky, like observing the crescent or half moon, looking at the double stars Albireo or Gamma Andromeda, or some of the brighter Messier objects?

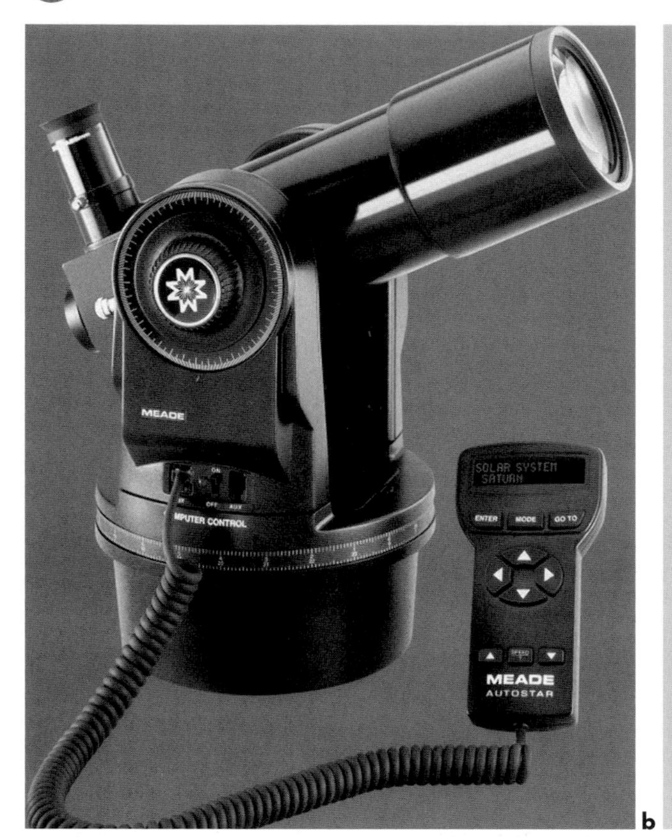

b

Figure 4.1. b The Meade ETX 70AT with standard controller. Photo: courtesy Meade Corporation.

The Cheapest Go To Instruments

Unlike the ETX 90EC, 105EC, and 125EC, Meade's 60-AT and 70-AT ETX telescopes are short-focus, basic-quality refractors working at f/5.8 and f/5.0, respectively. Their focal lengths are both 350 mm. They are classed as ETX telescopes because they share the same base, fork, and drive components.

That said, it doesn't take a genius to work out that compared to the f/13.8–f/15 focal ratios of the ETX Maksutovs, the smaller ETX refractors have focal lengths four or five times shorter and hence four or five times wider fields of view (for the same eyepiece). The maximum fields of view of these little refractors (about four degrees) are closer to those of the finders on the ETX Maksutovs. With a four-degree field, hitting the target is far more likely and you can dispense with a finder, saving critical dollars in the final production cost.

Meade's DS-2000 Digital Series telescopes don't use the ETX base, but do have similar slewing capabilities with Meade's Autostar controller. The DS-2000 series

Figure 4.2. The author's Meade ETX on the BC&F ETX Tri-Wedge.

includes, at present, a 70-mm, f/10 refractor and two short-focus reflectors of 114 and 127-mm aperture.

Celestron's NexStar 80GT competes most closely with the small Meade refractors but has a larger 80-mm aperture with a 400-mm focal length. Like all four of the entry-level Celestron NexStars, a useful "zero power" finder is included. (The smallest Meades have no finders). In terms of overall beginner satisfaction the 80GT is probably the best of Celestron's lowest-price offerings because its aperture is useful and its field of view is compatible with the slewing accuracy of the simple drives.

In contrast, the NexStar 60GT is a 60-mm, f/12 refractor with a small aperture and 720-mm focal length, and the 114GT is a 1026-mm focal length Newtonian reflector.

Given a choice of these three, I'd personally choose the 80GT for user-friendliness every time. The NexStar 4GT is a slightly different offering; a cut down version of the larger NexStars, featuring the same single-arm support, but with 100-mm aperture, f/13 Maksutov optics.

LX90 and LX200

Having talked about Meade's smallest slewing telescopes let's move up a size, to Meade's LX90 (see Figure 4.3). The LX90 was designed to specifically rival Celestron's NexStar 8, a telescope whose smaller brother (the NexStar 5) was itself designed to upstage Meade's ETX series! The LX90 features the ETX Autostar controller (in my opinion a more capable unit than the original LX200 controller) but cuts costs by having a slightly smaller drive system than the LX200. Crucially, the LX90 has a larger worm and wheel gear than the ETX, so the Autostar controller's impressive features can be more fully utilized.

So does the LX90 deliver the goods when it comes to slewing accuracy and tracking accurately enough to take good photographs and CCD images? The answer is "yes."

The compromises made to reduce the cost of the LX90 (compared to the LX200 range) have been to omit the periodic error correction (PEC) facilities and to design a smaller and lighter fork and drive motor assembly. None of this compromises the slewing accuracy and purchasing an autoguider or stacking short exposures can overcome the loss of PEC.

Figure 4.3. Leeds Astronomical Society members Robin Carmichael and Xavier Vermeren (at the eyepiece) with Robin's Meade LX90. Photo: courtesy Ray Emery.

The standard LX200 drive components can cope with LX200s up to 14 inches (35 cm) in aperture, whereas the LX90 is designed specifically for the 8-inch telescope. From an electronic engineer's viewpoint, some of the components used in the LX200 have been looking old-fashioned for years, so a major design change (the Autostar system) was probably overdue. The new system, first introduced on the ETXs and then moved to the LX90, has now been fitted to the new LX200 GPS (Autostar II).

The current Autostar features include:

1. Go To for 13,235 deep-sky objects, i.e., all the NGC, IC, Messier, and Caldwell objects.
2. Go To for 16,888 stars, all the planets plus the Moon
3. Go To for 26 asteroids, 15 comets, 50 Earth satellites, and 200 user-defined memory locations.
4. Guided tours of showpiece objects and special event menus (e.g., eclipses, minima of Algol, sunrise/moonrise and sunset/moonset data, and times of meteor showers).

The LX90 is incredible value for money.

So where does the LX200 GPS (see Figures 4.4a. and 4.4b.) stand in comparison with it? As far as the 8-inch (20-cm) LX200 GPS is concerned, the only difference is that the LX200 GPS has a better drive, equipped with PEC and the more powerful Autostar II control system and database (with 145,000 celestial objects).

Although PEC is not available on the LX90, a CCD autoguider can be used to correct the drive errors, provided you purchase Meade's optional $50 accessory port module. Many leading amateur CCD imagers prefer autoguiding to PEC for improving tracking for long exposures. For short, say 60-second, exposures such as you might need for supernova patrolling, the PEC capabilities of the LX200 are probably easier.

If you are *really* keen to take the best images, you may want to purchase SBIG's AO7 adaptive optics autoguider, which I'll discuss later!

The LX200 GPS with its 145,000-object database slews even more accurately to targets than the LX90 and, more importantly, is available in 8, 10, 12, and 14-inch (20, 25, 30, and 35-cm) apertures. With the larger models it is essential to balance the telescopes properly in RA and Dec, especially when they are equatorially mounted – the drive systems are under a lot of stress when slewing at 6 degrees a second. Meade's flagship 16-inch (0.4-m) LX200 is in a different category – more about that later.

The GPS feature on both Celestron and Meade's telescopes is, for me, something of a gimmick. All it means is that on switching the telescope on, it can use the GPS satellites to work out where it is, usually within a few tens of meters. The telescope also knows the exact time without the user having to key it in from a radio-controlled clock – that's a definite asset. It certainly saves keypad fumbling in the dark.

There are other integrated features included, to do with finding north and levelling the telescope base, but it's still, to me, a gimmick. Others will no doubt disagree.

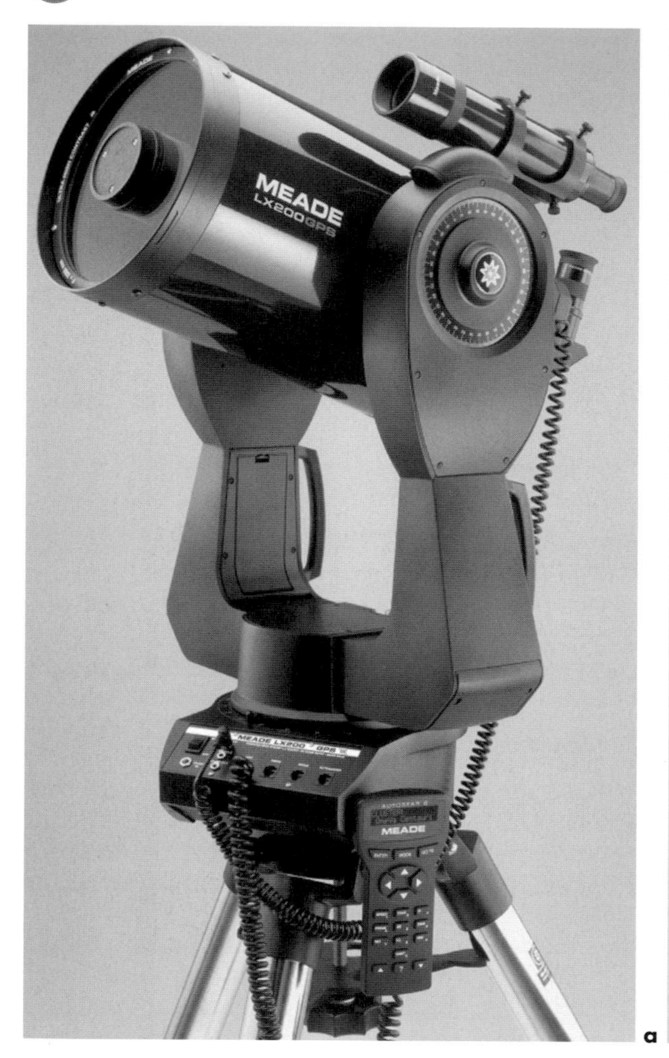

Figure 4.4. a The 8-inch (20-cm) Meade LX200 GPS. Photo: courtesy Meade Corporation.

Celestron

Celestron's response to the launch of the 90-mm ETX and its 127-mm big brother was the NexStar range, starting with the NexStar 5, a head-to-head rival for the bigger Meade instrument. The NexStar design has proved so popular that it has swept upward in aperture through the Celestron product range, offering 20- and 28-cm models, with a GPS location system option. Although the NexStar 5i (Figure 4.5) and 127-mm Meade ETX have identical apertures, optional computerized slewing, and similar features, there are differences, namely:

Figure 4.4. b The 14-inch (35-cm) Meade LX200 GPS. Photo: courtesy Meade Corporation.

- The NexStar 5i is an f/10.8 Schmidt–Cassegrain giving a wider field of view than the ETX, but without the fixed collimation of the f/13.8 Meade Maksutov.
- The NexStar 5i has a zero-power finder for easy alignment on naked-eye stars whereas the ETX has an 8 × 25 finder with a 5-degree field of view.
- The NexStar 5i has keypad buttons dedicated to Messier, Caldwell, and NGC objects whereas the ETX requires menu scrolling to select targets.
- In tests I've carried out the NexStar 5 hit the targets more often than the ETX, even allowing for the NexStar's wider field of view with the same eyepiece.
- The ETX has clutches on each axis so experienced observers can forget the motorized slewing and position the scope manually.

Figure 4.5. The Celestron NexStar 5i. Photo: courtesy Celestron International.

The other noteworthy feature of the NexStar 5i is its dramatic-looking single-arm design, in contrast to the traditional twin-arm fork used for Schmidt–Cassegrains for the past 30 years. Only one volume manufacturer, Quantum, had used this design in the past, on its Maksutov rival to the legendary Questar Maksutovs.

The NexStar fork arm also has a nice touch, in that the telescope hand controller can be tucked away, and used, in the fork arm. If the NexStar 5i or Meade 125EC were much heavier, lugging them outdoors would be less than enjoyable. However, if your muscles are larger than mine (not difficult) you may be happy transporting the larger telescopes in the NexStar range.

I should stress that the Celestron range of telescopes is considerably larger than I am portraying here, and is constantly changing and evolving. This makes it almost impossible for a writer to be specific because by the time you read this book the specifications may well have changed. Manufacturers' Web pages are the best source for the latest information (see the Appendix).

The NexStar 8i/8GPS Versus the Meade LX90/LX200GPS

Celestron's NexStar 8i and 8GPS, not surprisingly, compete head-to-head with the Meade LX90 and 20-cm LX200 GPS. It's the ETX 127 versus the NexStar 5 all over again. Or is it? All these scopes are 20-cm, f/10 Schmidt–Cassegrains with similar central obstructions.

Essentially, Celestron's NexStar 8i (Figure 4.6) is a "budget aperture solution": a bigger tube on the 5i mount, with optional hand controller and tripod. It's primarily aimed at the visual observer. The NexStar 8 GPS comes with a heavy-duty tripod and has high-quality drive components and a low-expansion carbon-fibre tube. It competes directly with Meade's 20-cm LX200 GPS, it has PEC, and an autoguider port. It's ideal for entry-level CCD imaging. Both the NexStar and Meade GPS scopes cost around $2500 at the time of writing.

Meade's LX90 is a $1700 instrument for visual observers with CCD aspirations, but as we've seen, there's no PEC and autoguiding is an optional add-on. The NexStar 8i is the budget $1200 single fork arm visual instrument, with the option of a computerized hand controller and GPS alignment options; it will slew an object into a low-power eyepiece field almost every time, but not on to the small field of a CCD chip.

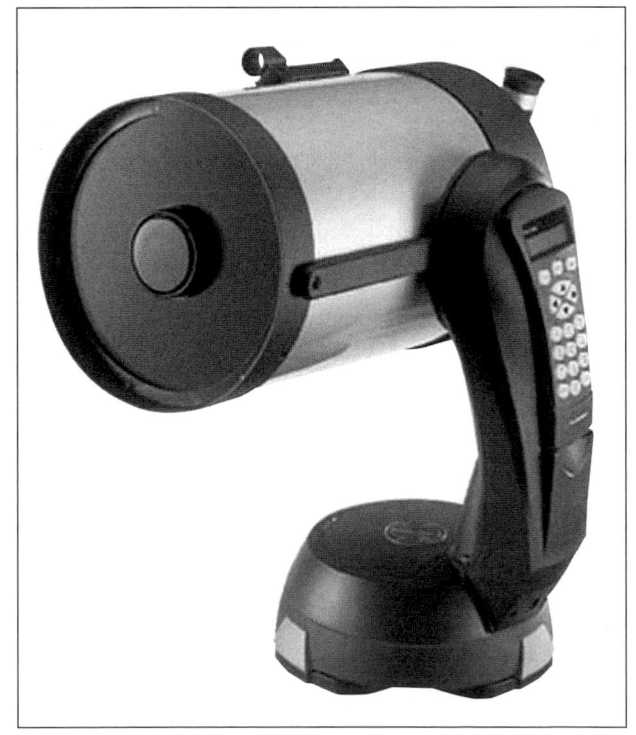

Figure 4.6. The Celestron Nexstar 8i. Photo: courtesy Celestron International.

From a portability point of view, the NexStar 8i is very light (24 lbs compared to 30 lbs for the LX90) and its tripod is lightweight too. The LX90 does have carrying handles on the fork tines, but most people will feel distinctly insecure lugging either instrument about in the dark unless they have a friend to help.

The NexStar 8i has the same single arm support and base as the NexStar 5i, but the 8GPS has a double arm fork. The longer tube means that the 8i can't be used to observe objects at the zenith. As with the NexStar 5i, the 8i has no setting circles (the Meades do) and it is intended to be slewed with the motors, whereas the LX90's dual clutches allow manual pointing. The 8i has the same zero-power finder as the NexStar 5i, whereas the 8GPS, LX90, and LX200GPS instruments have quality 8 × 50 finders which are nice accessories to have for confirming the field. For observers in many locations, a clear night often means a frosty night, and the NexStar series LCD displays can sometimes become tricky to read in the dark, when it's freezing cold.

From a power point of view, the much bigger battery compartment of the LX90 takes eight C cells, but the NexStar 8i only takes eight AA cells. Bearing in mind that the Celestron has to be slewed by the motors, one new set of batteries may well not last even one freezing-cold night session (four hours is a reasonable estimate under normal usage). At least the NexStar battery compartment is easily accessible! Celestron also supply a mains power unit for those within reach of a mains socket.

The LX90 has 30,000 Solar System, Deep Sky, Stellar and User-Defined objects; the NexStar 8i database (optional extra) and NexStar 8GPS have 40,000, and the LX200 GPS has 125,000. I do like the Celestron controller's dedicated Messier, NGC, Caldwell, Planet, and Star buttons though (see Figure 4.7).

The 25–35 cm SCTs

Excluding Meade's white-tubed giant 40-cm LX200, the LX200 GPS range stretches all the way from 20 to 35 cm. In the early 1990s there were just 20- and 25-cm LX200s. Then a 30-cm was added to the same fork mounting (I own one!). Then, in late 2002, a 35-cm tube was added to the same (gulp!) mounting. Okay, the fork tines were wider (obviously) but it was, essentially, the same mounting. The 25-, 30- and 35-cm LX200 GPS models share all the characteristics of the 20-cm LX200 GPS mentioned above, they just have bigger optics.

Celestron's NexStar range has just one extra model, the 28-cm NexStar GPS (Figure 4.8) and at $3000 it competes directly with Meade's $3000 25-cm model. With the introduction of Meade's 35-cm LX200 GPS at $4300 and with their 30-cm LX200 GPS dropping to $3500, competition is intense at the large-aperture end of the market!

All these instruments are excellent for visual or CCD use and are all superb technology for the money. I like the look of Celestron's motor drive gear (both axes feature high-quality 143-mm diameter, 180-tooth worm wheels); I also like Meade's LX200 mirror lock feature, to prevent focus-shift when slewing. The carbon-fibre Celestron NexStar tube is a bonus, as are the Fastar compatible

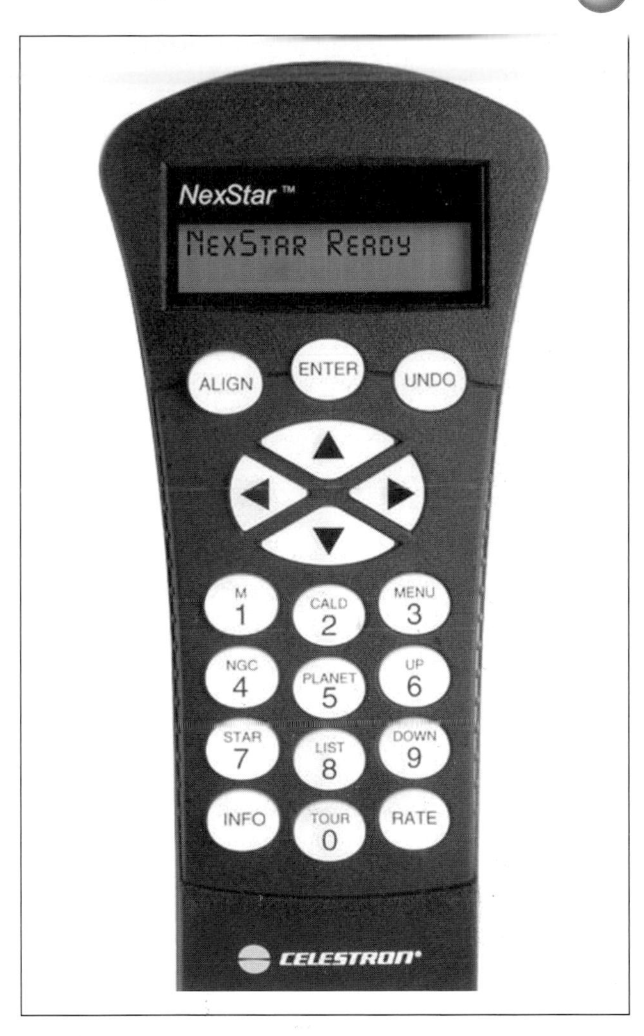

Figure 4.7. The Celestron NexStar Controller. Note the dedicated Messier, NGC, and Caldwell keys. Photo: courtesy Celestron International.

f/2 optics (you can place a CCD camera at the Celestron's NexStar GPS's f/2 prime focus for wide-field CCD imaging).

Meade's Go To LXD55 Telescopes

In 2001, Meade introduced yet another product range based around a modest German equatorial head. The product range was designated LXD55 and featured three optical configurations, namely:

- Achromatic refractors of 5- and 6-inch (127- and 152-mm) aperture;
- An 8-inch (203-mm) Schmidt–Cassegrain;

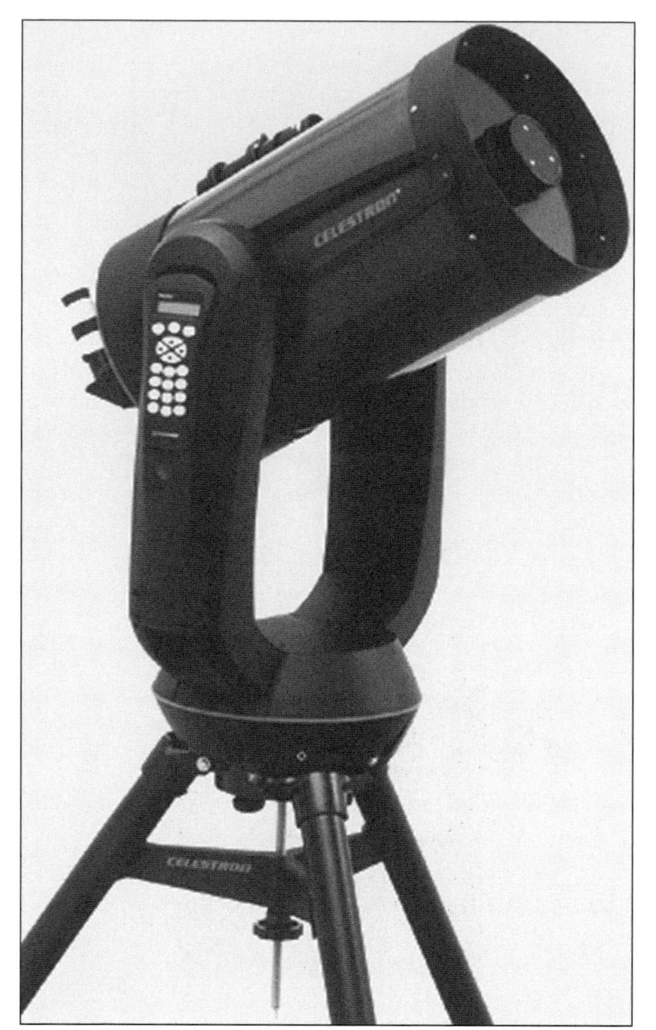

Figure 4.8. The superb Celestron Nexstar 11 GPS. Photo: courtesy Celestron International.

- A new range of 6-, 8-, and 10-inch (152-, 203-, and 254-mm) Schmidt–Newtonians (i.e., Newtonians with an aberration-correcting glass plate and closed tube).

It is the Schmidt–Newtonians that have generated the most interest as they really are astonishing value for money.

These telescopes are best thought of as visual instruments, with the drive system being better in quality than the basic ETX, but certainly not in the LX90/LX200 league. With the Autostar option, the ideal optics to go for would be the wide-field Schmidt–Newtonian tube assemblies. The low prices of these instruments are quite remarkable, for example: $700 for a 6-inch, f/5 Schmidt–Newtonian with Autostar and a 30,000 object database; $1000 for a 10-inch, f/4 Schmidt–Newtonian with Autostar. The former instrument, used with a 26-mm eyepiece gives 29× magnification and a 1.7-degree field. The combination of a large aperture and a wide field, plus a better drive, makes the $700 instrument a better prospect for the beginner than the $650 ETX 90EC with Autostar option.

Such an instrument would not be as portable as an ETX, of course, but is probably more satisfying. The LXD55 range is not really intended as a CCD imager's telescope, but as a value-for-money beginner's visual telescope, it really has no rivals. A nice touch with thc Autostar models is that if the Go To slew misses the target, pressing Go To again searches around in a spiral until – with luck – the user can spot the object! Figure 4.9 shows a Meade LXD55 Schmidt–Newtonian.

Figure 4.9. Meade's 8-inch (20-cm) LXD55 Schmidt–Newtonian. Photo: courtesy Meade Corporation.

Ultimate Go To Telescopes

So far, we've been considering the sort of telescopes that a keen amateur might buy for a few thousand dollars. But what about the ultimate amateur telescopes – instruments costing between $10,000 and $20,000 dollars?

Probably the three best-known contenders are the 16-inch (0.4-m) Meade LX200 and the Software Bisque Paramount ME or Astrophysics AP1200GTO mounts with a Celestron 14 Optical Tube Assembly combination.

The 16-inch LX200

The 16-inch LX200 has been around since 1994 and is in a different category from both the smaller LX200 and the LX200 GPS models, even allowing for its larger aperture.

Gordon Rogers' 16-inch LX200 is shown in Figure 4.10a, and Gordon himself in Figure 4.10b. The 16-inch features a *much* more substantial fork mounting and drive system than the other Meades. The 16-inch LX200 contains 280-mm diameter worm wheels on each axis and precision 80-mm (Dec) and 100/150-mm

a

Figure 4.10. a Gordon Rogers's 16-inch LX200. Photo: courtesy of Gordon Rogers.

Figure 4.10.
b Gordon Rogers in his observatory. Photo: courtesy of Gordon Rogers.

(RA) roller bearings, necessary to take the weight, of course. The pointing precision is, typically, about 2 arc-minutes.

The 16-inch Meade also has a "home pulse" feature to maintain the telescope's pointing position when it is switched off (useful for remote operation) and a thermal stabilization fan in the rear cell to rapidly cool the optics to the night-time air temperature (very useful for planetary work). Despite the huge size of the 16-inch LX200 fork assembly the telescope is remarkably compact when used in alt-azimuth mode. For this configuration, Meade can supply their accessory #1222, a field de-rotator, which plugs into a dedicated socket on the telescope control panel. This gadget rotates the camera or CCD attached to the telescope to compensate for the fact that the field seen in the eyepiece rotates when a telescope is tracking in alt-azimuth mode. In practice the field-derotator is needed for exposures longer than about 2 minutes. If you plan to use a telescope to piggy-back other photographic equipment then you will always need to polar-align the scope.

However, for a telescope as large as the 16-inch LX200, the alt-azimuth configuration saves a *huge* amount of space – and, of course, money (the 16-inch equatorial pillar is massive – look at Figure 4.11!).

The Paramount ME + C14

While the 16-inch LX200 is an impressive "amateur" instrument, the Software Bisque Paramount mounting (Figure 4.12) is truly in a different league and, arguably, the best amateur telescope mounting in the world. It is, without doubt,

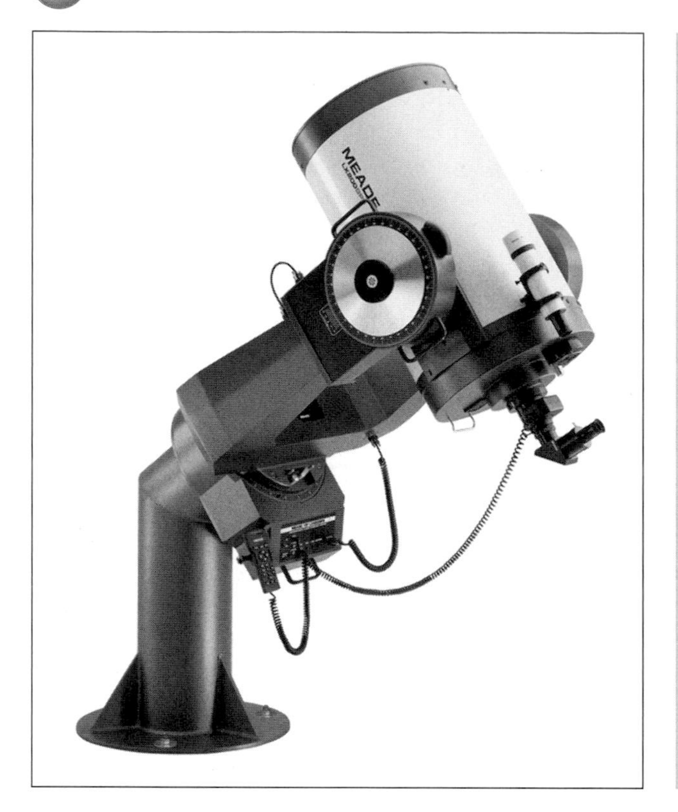

Figure 4.11. Meade's 16-inch (40-cm) LX200 on its equatorial pier. Photo: courtesy Meade Corporation.

Figure 4.12. Software Bisque's Paramount ME carrying a 16-inch (40-cm) Ritchey–Chretien Optical Tube Assembly. Photo: courtesy Brad Ehrhorn/RC Optical Systems.

the mounting of choice for supernova patrollers worldwide. The "dream combination" has always been the Paramount mounting mated to the Celestron 14 SCT Optical Tube Assembly, though with Meade now offering a 14-inch Optical Tube Assembly, there are two 14-inch options.

In 1997 Software Bisque decided to diversify into mechanical engineering, to produce the Paramount GT-1100 telescope mounting, a mounting that really would do justice to their telescope control software. Various improvements have been made to the original mount in the last five years; in late 2000 the GT-1100s was announced and, a year later, the "Millennium Edition" Paramount GT-1100 ME became available, for the relatively modest price of $8500 (now $10,000). It did not take long for Software Bisque's superb mountings to be in great demand from amateurs and professionals alike. The first four Paramount mountings were sold to the US Air Force for locating spy satellites. Many more were sold to keen amateurs who had been looking for this quality of mounting for some time. One of the first purchasers was the US amateur Michael Schwartz. Michael was really the first amateur to achieve consistent supernova discoveries by using the Paramount to patrol hundreds of galaxies each night. By combining the accurate slewing and tracking of the mounting with the large aperture of a 14-inch (356-mm) Celestron (working at f/11 for maximum penetration) and an ultra-sensitive back-illuminated CCD camera, supernova discoveries came thick and fast. At the time of writing he has discovered more than 70 supernovae (see Chapter 9).

So what design features make the Paramount mounting so good? First, the mechanical components are of the very highest quality: research-grade worm and wheel mountings on the RA (290-mm diameter) and Dec (180-mm diameter) axes are just the start. The gear-reducer and worm-block design are so efficient that hardly any power is lost between the motor and the telescope drive; in addition, the belt-driven research-grade gears have virtually no backlash. The RA ball-bearing assembly on the Paramount is a substantial 8 inches (20 cm) in diameter. With regard to the drive motors, the Paramount uses brushless DC servo motors. These are expensive, but the only moving parts in these motors are the bearings, which means the life expectancy of the motors is very long. The motors used in commercial SCT mountings will not last many years if supernova patrolling is undertaken! One only has to listen to a Paramount under full slew, and then a commercial SCT under full slew to appreciate the difference in quality.

The Paramount mounting has always been supplied with professional *TPoint* tracking software that makes the telescope slew more accurately. The latest software features *ProTrack*, which uses the *TPoint* modelling data to make the mount track better. Put simply, once you have eliminated any periodic error of the mounting using the second- to fifth-order polynomial curve-fitting software supplied, you are left with atmospheric refraction, tube flexure, polar misalignment, and gear eccentricity as the major sources of tracking error. While earlier versions of *TPoint* improved the Paramount slewing to sub-arc-minute precision, *ProTrack* uses the same model to improve the tracking. So dramatic are the improvements that four-minute *unguided* CCD exposures at 2-meter focal lengths are possible! The mount can also be programmed to follow comets and asteroids moving slowly against the background stars.

When a telescope can patrol a thousand galaxies a night, night after night, in summer or winter, and *never* suffer a breakdown and show no signs of wear, year after year, this is total reliability; this is what the Paramount can do. Perhaps the only disadvantage of this excellent mounting is its need to "normalize" when the German equatorial mounting causes the telescope tube to cross the meridian. (In other words, the tube and counterweight swap places to prevent the tube hitting the pier.) This is a minor point. But remember, we're talking $10,000 for this mounting.

Before we leave the realm of dreams, let's briefly have a look at some other superb telescope mountings.

Astrophysics 1200GTO and Takahashi Temma

An alternative high-quality mounting supplier is the Astrophysics Company, based in Rockford, Illinois. Astrophysics have long been renowned for their unsurpassed range of apochromat refractors, telescopes that deliver exquisite views of the night sky; however, their German equatorial mountings are equally impressive and for some years have boasted a novel feature – voice-actuated telescope control.

The flagship Astrophysics mounting is the 1200GTO mount which features a 26-cm, 225-tooth RA worm wheel and 18-cm, 225-tooth Dec worm wheel and substantial RA and Dec shafts and bearings. Payload capacity is 63.6 kg (140 lb). The drive can be controlled over a range of 0.25× sidereal to 1200× sidereal, i.e., 5 degrees per second slewing (on both axes). In addition, the weight of the mounting is only 41.4kg, or 91 lbs. The asteroid discoverer and astrometrist Stephen Laurie is one of many top observers who uses an Astrophysics 1200GTO, in his case to carry a Celestron 14 Optical Tube Assembly. The Astrophysics 1200GTO mount is shown in Figure 4.13. Astrophysics can supply *DigitalSkyVoice*™ software for use with this mounting. The software, which runs on a standard PC equipped with a microphone, recognizes key audio commands and slews to the target as directed by the user. So, no more fiddling about with keypads in the dark; but it does take a bit of practice to learn to use it well! Many experienced astrophotographers hold the opinion that, apart from professional mountings, costing $20,000 and above, the Paramount ME and Astrophysics 1200GTO are in a category of their own – i.e., the ultimate amateur mountings.

Takahashi is another name that has long been associated with quality and they now offer a quality mount that slews to different targets. The Takahashi "Temma" mountings are available in four different sizes (i.e., for Takahashi's EM-10, 200, NJP, and EM-500 mounts) and with slew rates of 150× sidereal or 400× sidereal, i.e., 0.6 or 1.7 degrees per second. These mountings are designed to carry payloads ranging from small refractors to 300-mm aperture Newtonians. It should be borne in mind that Takahashi are very conservative with their specifications: their mountings could easily take much heavier payloads than advertised! Also, the addition of a slew facility on their mountings has not compromised tracking accuracy; their mounts may slew more slowly than the mass-produced SCTs, but the tracking accuracy is excellent.

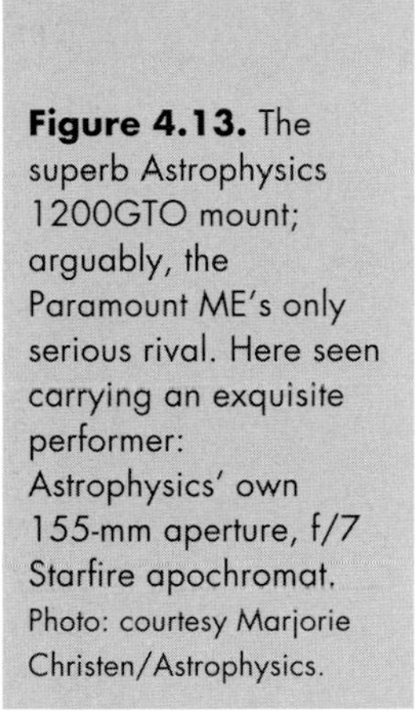

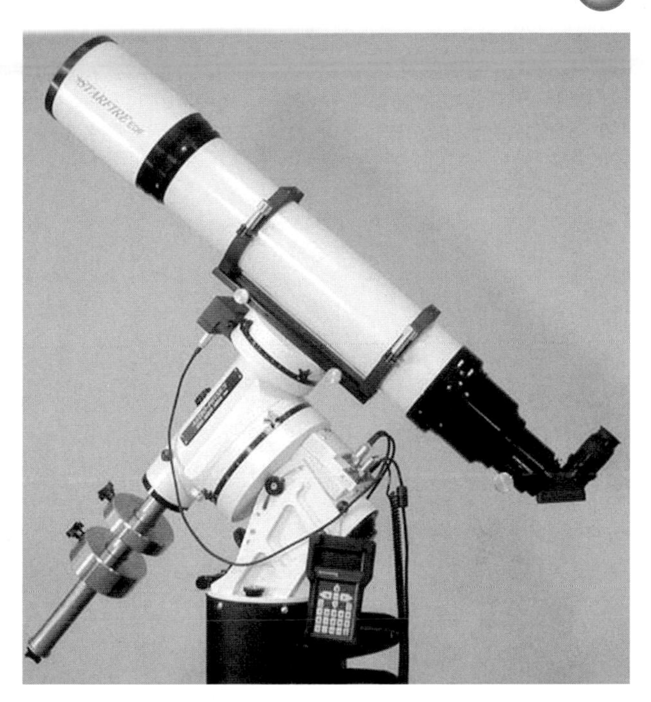

Figure 4.13. The superb Astrophysics 1200GTO mount; arguably, the Paramount ME's only serious rival. Here seen carrying an exquisite performer: Astrophysics' own 155-mm aperture, f/7 Starfire apochromat. Photo: courtesy Marjorie Christen/Astrophysics.

As with the Paramount, these mountings are not cheap, ranging from $3000 to $12,000 for amateur payloads and up to $80,000+ for professional instruments. I guess you get what you pay for.

As this book was going to press, Celestron were unveiling some quality telescope systems which may well fill a gap between the NexStar/LX200 class of instruments and instruments of the Paramount ME quality. Marketed under the CGE brand they feature German equatorial computer-controlled mountings for apertures from 20 to 35 cm. It will be interesting to see how these instruments compete with existing models in the $3000–6000 price range. Another interesting newcomer, just released at the end of 2003, is the Vixen Sphinx GO-TO mount. Intended for light payloads of up to 10 kilograms, it looks like an attractive, low-cost option for 8- or 10-inch (20 or 25 cm) instruments.

CHAPTER FIVE

Digicams and Video Astronomy

Digicams versus Cooled CCDs

Prior to 1990 the vast majority of amateur astrophotographers used photographic film to record objects in the night sky. Since the early 1990s there has been a gradual shift towards charge-coupled devices (CCDs), and with the possible exception of visual variable star and comet magnitude estimates, every aspect of astronomy has been transformed by the CCD detector.

Non-astronomers also enjoy the same benefits when using a digital camera (digicam) – but can such commercial equipment be used for astronomy?

There are some critical differences between astronomical CCD cameras and commercial digital cameras, namely:

- The CCD chips in astronomical cameras are cooled to about 30 °C below ambient temperature, which reduces thermal noise by a factor of 20.
- The lenses on all but the most expensive digital cameras cannot be removed and so interfacing to a telescope can require some ingenuity or accessories.
- There are far more pixels in the average digital camera CCD, so the image file size will be considerably larger (a factor when e-mailing or uploading images to the Web).
- Commercial digicam CCDs have color filters and use intensive image-processing chips, both of which are geared to optimize daylight results and dynamic range, not night-time sensitivity.
- The apertures (*not* the relative apertures, or f-ratio) of the lenses on compact non-SLR (single-lens reflex) digital cameras are tiny when compared to those on film cameras, especially when compared to the aperture of a standard

50-mm, f/1.8 lens. This is a direct consequence of the CCD covering about 1/20th of the area of a piece of 36 × 24 mm film. For the same field-of-view and the same f-ratio lens, the aperture (in area) will also be 1/20th of that of the film camera. Ironically, this more than cancels out the increased quantum efficiency of the CCD, so most pocket-sized digital cameras are no more sensitive than film cameras.

The astronomer who buys a digital camera for imaging star fields and constellations needs to ensure that the camera controls allow manual focusing, manual exposure, and ideally a B ("bulb") setting.

Most good digital cameras will have a "noise-reduction system" which acts like the dark frame in an astronomical CCD camera, subtracting the thermal noise pattern and allowing long exposures to be undertaken. If the maximum bulb setting is 1 minute or less, it is likely that the manufacturers have restricted the exposure because they know that longer exposures will be dominated by thermal noise in the CCD chip; indeed, the image may even be saturated by noise (i.e., the picture will be white!).

My own digicam, a Nikon Coolpix 5700, has an excellent 5-minute B (brief time) setting and is useful for taking tracked, wide-field constellation shots to magnitude 7, and deeper images at 8× zoom to magnitude 11.

When looking for digital cameras for astronomy it is essential to remember that a bigger lens aperture means more light and fainter stars. By dividing the focal length by the f-ratio, you get the aperture, but be careful to compare like-with-like; some manufacturers' data sheets will quote the equivalent 35-mm film focal lengths, e.g., they will quote 35–105 mm whereas the real focal length is only 9–27 mm.

Consider the advantages of using the camera at full zoom: the field of view will be much smaller, but the aperture is usually bigger, so more stars will be captured. The largest apertures available are those achievable with the "professional SLR digital cameras", i.e., those cameras that have removable lenses and large CCD chips approaching the size of 35-mm film frames, and are intended to be used with standard 35-mm camera lenses. Examples are the Nikon D1, Canon D60, and Fujicolor Finepix S2 Pro. These cameras can invariably use the large-aperture lenses designed for the manufacturers' own 35-mm film cameras and this can produce a significant improvement in limiting stellar magnitude. Typically, a 30-second exposure with such a digicam and a 50-mm, f/1.8 lens will reach mag 10, whereas a standard 8- or 9-mm focal-length digicam image will reach mag 6 or 7. Such SLR digicams, with removable lenses, can be used like film cameras, to take deep-sky images through a telescope. Their sensitivities are midway between those of film and a dedicated, cooled CCD camera, and stacking loads of short exposures on top of one another can drastically reduce the noise in the individual images.

As this book was in the final stages of preparation, Canon amazed the digital camera world by introducing two remarkably low-priced digital SLRs, which look set to become the amateur astronomers digicams of choice. Priced at around the $1,000 level, the Canon Digital Rebel (US) or 300D (UK) is fast entering every amateur's wishlist. Its more expensive counterpart, the Canon EOS 10D, is already proving itself a worthy contender for non-cooled CCD imaging. The situation in this exciting area is changing month by month.

Perhaps the best astronomical roles for a commercial digicam are the fields of bright nova patrolling, meteor photography, and aurora photography. However, resourceful amateurs have also used digicams for lunar, planetary, and solar imaging. Let's look at nova patrolling first.

Nova Patrolling

Figure 5.1 shows the prime nova hunting regions available to the observer in north temperate latitudes in the spring, summer and autumn months. It depicts 26 confirmed novae brighter than mag 9, in the summer Milky Way, above Dec −20 but below Dec +55 and between 1967 and early 2001. It also shows four very bright novae in the same region from 1918, 1920, 1934, and 1963. Observers further south enjoy the teeming nova fields of the central Milky Way, of course, i.e., Sagittarius and its neighboring constellations.

The box is 30 degrees wide by 40 degrees high, and has corners at Right Ascension/Declination points of 20h −10 and 18h +30. The boxed region is suggested as a good, but manageable search area for a digital patrol. (A 50-mm focal-length lens with 35-mm format film would cover a slightly smaller area.)

A modern SLR digicam could easily patrol the boxed region to mag 10, with just a few exposures of 1 minute or so duration.

Another small-aperture solution is to use a standard film camera fish-eye lens coupled to an astronomical, cooled, CCD camera. The (typically) 16-mm focal length recovers the field lost by the small chip size. Though not a digicam solution, this is an effective CCD nova patrol solution and one that I use myself (Figure 5.2 shows the northern Aquila and Vulpecula region imaged with a 16-mm, f/2.8 fish eye and an MX916 CCD.)

Total coverage of the boxed region could have netted 17 novae in the last 34 years, i.e., one every two years; and that's only the ones we know about!

Okay, we've established that nova patrolling can be carried out with digital cameras (and CCDs attached to fish-eye lenses) so what else can we do with them?

Lunar, Planetary, and Solar Imaging with a Digital Camera

Figure 5.3 shows an image of a lunar occultation of Saturn taken by Hazel McGee in her first ever attempt at digital astrophotography; proof, if any were needed, that low-resolution lunar imaging is relatively trouble-free. In this instance the camera was simply pointed down the eyepiece. As the vast majority of digital cameras have non-removable zoom lenses, this is one of the few methods of achieving a result.

The problems with pointing a camera down the eyepiece (afocal imaging) are twofold, namely: holding the camera steady, and conveying all the light from the telescope eyepiece to the CCD chip without cutting off or vignetting the light.

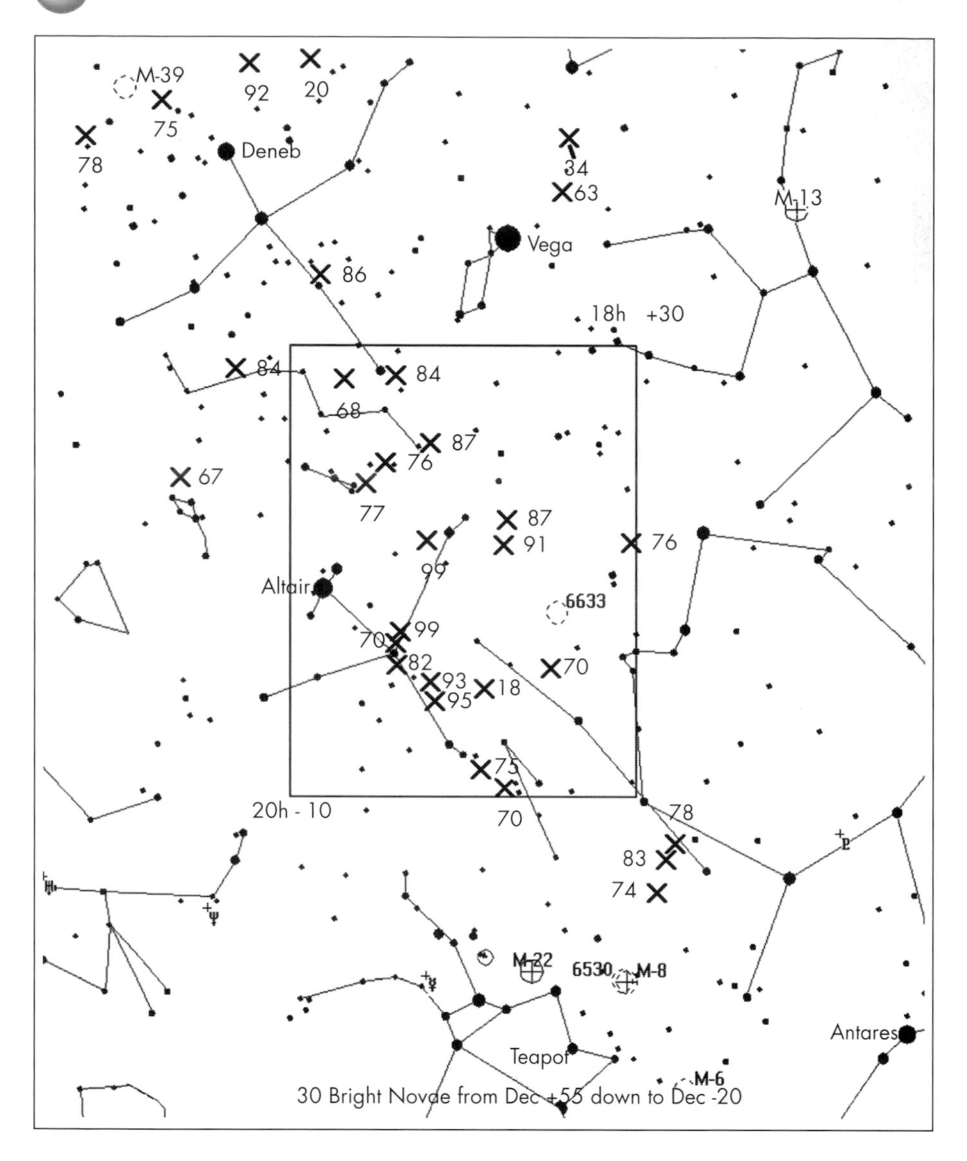

Figure 5.1. The prime nova patrol region for northern hemisphere observers. The boxed region is 30 × 40 degrees.

The first problem is easily solved; you can now buy digital camera adapters for telescopes from astro-accessory stores. Many dealers sell interfacing gadgets to help form a rigid coupling between eyepiece and camera.

The second problem is trickier and more dependent on the size of the camera lens with respect to the eyepiece lens and the precise eyepiece design. At first glance it might be thought that the priority here would be to have a large camera

Figure 5.2. Northern Aquila to Vulpecula/S Cygnus, taken with a 16-mm, f/2.8 fish-eye lens attached to a Starlight Xpress MX916 CCD. A three-minute exposure by the author.

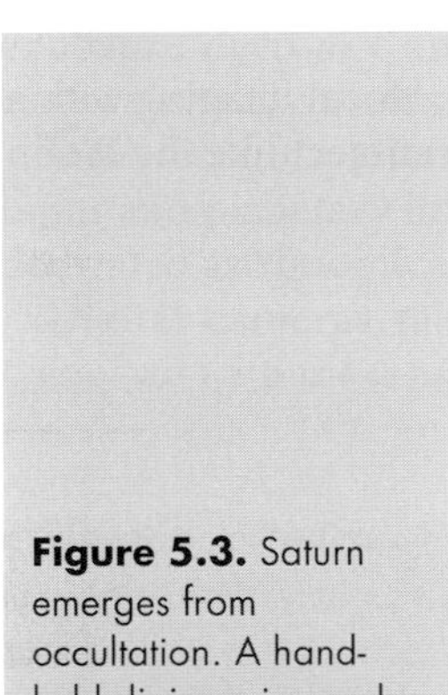

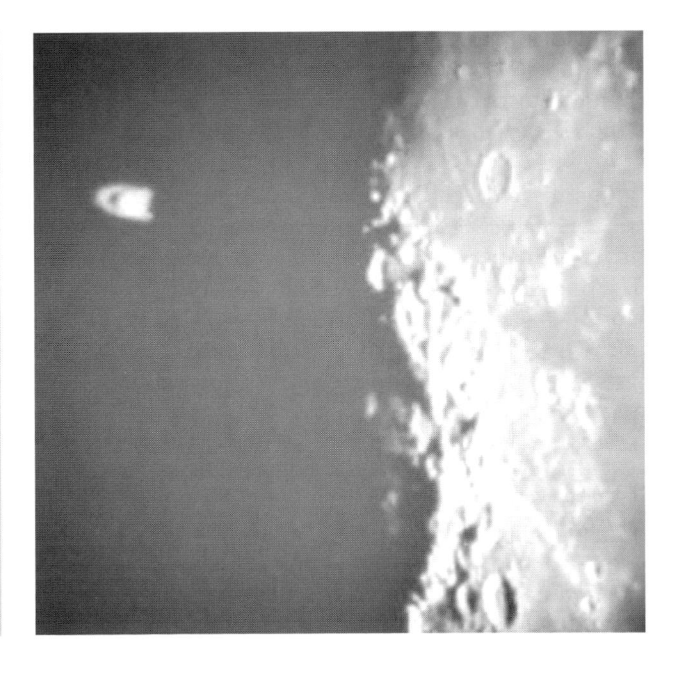

Figure 5.3. Saturn emerges from occultation. A hand-held digicam image by Hazel McGee. The camera was simply pointed down the eyepiece of a 30-cm LX200. 3 November 2001. Photo: courtesy of Hazel McGee.

Figure 5.6. a Ray Emery's 80-mm, f/5 Celestron refractor plus 40-mm aperture Coronado hydrogen-alpha filter. **b** Ray Emery pointing his Coolpix 885 digital camera into the eyepiece of his 80-mm refractor. Photos: courtesy Ray Emery/John Ericson.

a thousand per night, at a very short exposure (so the seeing is frozen and I can catch those rare moments of near-perfection). The image on the PC monitor is very dim and I can't tell if it's an excellent image or not; I just save it anyway. The next day I wade through the thousand or so images in a dark room, looking for the best ones. With the limited storage capacity of digital cameras this can be a restriction, unless a 1 Gb microdrive or a very large flash card is used.

Solar imaging is another field that is especially suited to digital imaging, especially with small telescopes, such as 60–80 mm aperture refractors.

Of course, as with all solar work, **never** look at the Sun through a telescope, unless you are 100 percent confident that a high-quality full-aperture solar filter is being used. The same applies to digital imaging of course – full-aperture solar filters must be employed at all times. The Sun is so bright that even with solar filters there is never any need for long exposures and rarely any need for large apertures. The atmospheric seeing is usually so poor in the daytime that a 40–60 mm aperture can often show all of the detail available. Using a small aperture is quite an advantage in afocal solar work because with, say, a 60-mm refractor, a low eyepiece magnification image of the Sun, spanning an apparent field of 10 degrees or so, will easily fit on the camera's CCD chip, well inside the vignetted region. This means that a clean image of the whole solar disk can be obtained. With the number of pixels now available in digital cameras, the magnification can be increased to yield one CCD pixel for every millionth of the solar hemisphere, a useful unit in solar observing!

In recent years, amateurs equipped with digicams have been taking superb images of the Sun's prominences using H-alpha filters attached to small telescopes. Ray Emery, of the Leeds Astronomical Society, has been a UK pioneer in this field. His equipment is shown in Figures 5.6a and b. Ray uses a Celestron 80-mm, f/5 short-focus refractor equipped with a Coronado H-alpha narrowband filter of only 40-mm aperture. Amazingly, he *hand holds* his Nikon Coolpix 950 at the eyepiece of the system, but still obtains stunning results on a weekly (and sometimes daily) basis (see Figure 5.7).

As well as using digital cameras for astronomy, many amateurs are finding that modern camcorders, security cameras, and webcams have a role in imaging the night sky, too, and we will look at this aspect of amateur astronomy next.

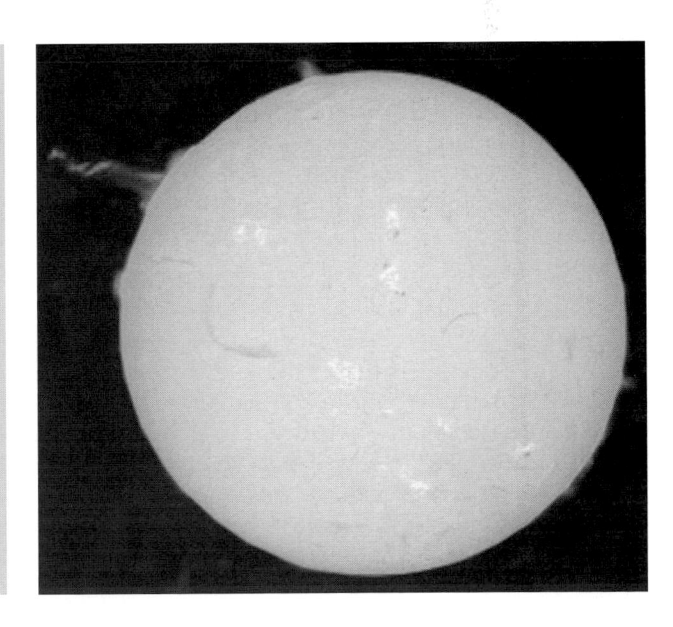

Figure 5.7. Hydrogen-alpha image of the Sun by Ray Emery on 26 April 2002, using the equipment shown in Figure 5.6. Celestron 80-mm, f/5 plus 40-mm Coronado H-alpha filter. Nikon Coolpix 885 on full zoom (22-mm, f/4.9) Digicam on 200 ISO setting. 1/60th sec.

Video Astronomy

Astronomy is not generally thought of as a hobby where things happen quickly but, when they do, the hobby is at its most dramatic. It is in these cases that domestic and advanced video equipment can play a part. Video also has a role in situations where, due to poor seeing, it's advantageous to take hundreds of very short exposures and stack the best shots up, rather than just take a few much longer exposures. The following subjects are ones which benefit greatly from video techniques:

- Medium-resolution planetary imaging;
- Total solar eclipses;
- Grazing lunar occultations;
- Planetary occultations;
- Occultations of bright stars by asteroids;
- Meteor showers (image intensifier required!);
- Freeze-framing of the Space Shuttle/ISS through a fast-slewing telescope.

Additionally, the proliferation of techniques for displaying video on a PC has encouraged many advanced amateurs to produce entertaining "speeded-up" animations of events that normally take place over many hours: I am thinking of animations of comets and asteroids moving across the sky, Jupiter rotating, and the lengthening of mountain-peak shadows near to the lunar terminator. Commercial software packages such as Richard Berry and James Burnell's *AIP* will align stacks of images of this type and convert them into an entertaining video with little fuss. With a bit more effort, animated JPEG or animated GIF routines can produce similar results.

Let's begin by having a look at the current equipment and techniques available.

Video Technology for Astronomy; Adirondack, Webcams, and Security Cameras

In the last few years, there have been a number of breakthroughs in converting webcams and security surveillance cameras for use in astronomy. Modern digital video (DV) camcorders have been sold with software and hardware packages to enable video frames to be captured and edited on a (new-ish) PC. In addition, inexpensive frame-grabbers, such as the *Snappy* or *SnapMagic* models have enabled standard video frames to be grabbed and imported to older PCs via the parallel port.

There are advantages to all this technology, even if the choices get more bewildering!

For the observer with a camcorder or DV, it is now relatively inexpensive to grab single frames for use in astronomy. Of course, single frames may not go that deep, but they are fine for shots of the Moon and planets. For many years (I was doing it in 1985) lunar observers have been videotaping the lunar surface and noting how similar the view was to the eyepiece view (and far superior to the photographic view!).

In recent years two bottlenecks in the astrovideo process have disappeared. These were first, the problem of storing many minutes' worth of 25 or 30 video frames per second on a PC, and second the processing time – the time taken for the system to grab each frame, store it, and being ready for the next one to arrive.

The first problem goes away with a multigigabyte hard disk, and the second has disappeared with the arrival of the latest generation of fast processors and superfast interface electronics like IEEE-1394 (Firewire). The Moon and planets can now be captured and stored at video rates, for hours on end if necessary! The advantage of this is that the moments of best seeing can be selected and numerous "best frames" can be stacked up to reduce noise. This is similar to the methods used, at much slower download rates, by the world's leading planetary imagers, using dedicated astronomical CCD cameras. "Stacking" techniques are discussed in detail in Chapter 8.

Dedicated astronomical cameras are, in most cases, less noisy (especially when cooled) and more quantum-efficient than video cameras, webcams, and security cameras. However, even with USB, they tend to download images every few seconds, not at 25–30 frames per second, so there are more images to stack with video, even if they are much noisier.

Webcams and security cameras can be modified to produce more gain and to integrate over several seconds, or minutes, instead of at the video rate. There are a number of Web pages for astronomy webcam enthusiasts; perhaps the best user group at the time of writing this is Dr Steve Wainwright's QCUIAG (QuickCam and Unconventional Imaging Astronomy Group) which is listed in the Appendix. Webcams and security cameras use CCD chips just like dedicated astronomical cameras, but some of the CMOS (Complementary Metal Oxide Semiconductor, if you must know) chips are quite noisy and, of course, webcams don't have features like autoguider chips that can interface to telescope drives!

Lack of chip cooling is a major problem if you use webcams for long exposures, as thermal noise will tend to dominate the exposure. But, once again, stacking images can partially compensate for this. A stack of 16 images will have a quarter of the random noise level of a single image of the same total exposure. Stacking can be taken to the ultimate extreme with the "drift scan" technique, where short exposures from a fixed camera or telescope can be downloaded at video rates and then stacked to allow for the Earth's rotation.

Specifications

When reading the specifications of camcorders you will see the low-light sensitivity expressed in *lux*, whereas in astronomy CCD cameras are rated by *quantum efficiency* and *read-out noise*. Amateur astronomers tend to prefer specifications which state the faintest star brightness (stellar magnitude) that can be recorded with a given telescope in 60 seconds.

The lux is the lighting unit for photovisual radiation (centered on 550 nm). But converting from this unit (1 lux is equal to an illumination power flux density of 1.47×10^{-7} W/cm^2!) to stellar magnitude is fraught with problems. Lighting engineers use a tungsten filament at a precise temperature of 2854 K and a visual-band filter (centered on 550 nm) to define the illumination level of one lux. The color temperature of a tungsten lamp clearly has little relevance to astronomy; nevertheless lux values are still quoted for low-light astronomical cameras.

Let's look at a few examples. The best modern camcorders can image objects outdoors in full moonlight; this level of illumination corresponds to a sensitivity of about 0.2–0.3 lux. If you attach a camcorder of 0.2–0.3 lux sensitivity to a 25-cm aperture telescope you should be able to image stars of about magnitude 5. The most sensitive high-gain, low-noise, monochrome security cameras can be hundreds of times as sensitive as this, without resorting to long exposures. With sensitivities of typically 0.005–0.0005 lux, a video camera can image stars down to 12th magnitude through a 25-cm telescope, with, typically, a 1/60th second exposure.

Commercial Systems

Probably the leading pioneer in commercial video astronomy equipment is Adirondack Video Astronomy of Glens Falls, NY, USA. At the time of writing they manufacture two types of tiny video cameras which are of great interest to amateur astronomers. Firstly, their Color PlanetCam is a $495 miniature color video camera. This camera is only five centimeters in width, height, and depth and weighs under 200 grams. The exposure can be adjusted from 1/60th to 1/1000th of a second and the video output can be recorded to VHS video or less noisy video formats like MiniDV. Tiny knobs and switches on the camera allow the user to alter the contrast, image sharpness, gain, and color balance.

This is not a deep-sky camera, however; it only works on the Moon, planets, or a solar-filtered Sun as it has a sensitivity of 1 lux. It will give very pleasing planetary images at reasonable image scales because the CCD chip has tiny 5-micron pixels. Unlike a conventional video camera it slots right in where an eyepiece would go: there is no lens.

Adirondack's second and third offerings are their amazing monochrome Astrovid StellaCam and StellaCam-Ex models. When I first saw the latter model advertised I thought it was breaking the laws of physics, as various users claimed they were imaging the central star in the ring nebula (M57) through small-aperture telescopes, with what was, essentially, a video camera! This star is sometimes quoted as having a visual magnitude of about 15.4 with most observers needing a 0.4-m scope to see it. However, its photographic magnitude is about 14 and what I didn't realize was that the StellaCam/StellaCam-Ex models can integrate up to 128 1/60th second frames giving an effective exposure of 2 seconds.

Ignoring noise considerations, this equates to five stellar magnitudes more sensitivity than a single frame. Now you can see how the camera can pick up such a faint star. Effectively, it is a bridge between a video camera and a long-exposure CCD camera.

It's not the only bridge: SBIG's STV is a more sophisticated system, with onboard storage, optional built-in monitor, and exposures from 0.001 seconds to 10 minutes. But the StellaCam/StellaCam-Ex cameras only cost $595/$695, whereas the STV costs from $1850.

Adirondack Video Astronomy (AVA) rate the sensitivity of the StellaCam and Stella-Cam-Ex as roughly 0.0001 and 0.00005 lux for a 128-frame stack.

The British Astronomical Association's Andrew Elliott, a pioneer user of video equipment for occultation work, has for some time used a low-priced video camera made by Watec – their 902H model. This camera has a claimed sensitivity of 0.0003 lux (which doesn't mean much unless it is qualified in some way: used with a 25-cm, f/10 Schmidt–Cassegrain, stars of magnitude 12 can easily be recorded on video). In conjunction with a fast, 50-mm, f/1.8 camera lens it will just reach magnitude 9.

The Watec 902H is not the only low-priced low-light camera suitable for occultation work; there are even cheaper cameras and the prices of all the models are dropping all the time. In 2002, the Watec 902H cost around $300.

The Supercircuits PC164C is only $130 if you can cope with half the sensitivity. This camera uses the 1/3-inch Exview HAD Sony chip which is also used in the CCTV Model 2006X camera. The CCTV camera is, I'm happy to say, very low priced and easily available in Europe. The 2002 price in the UK was a mere £69 + tax from RF Concepts of Dundonald, Northern Ireland.

Meade also make an inexpensive "electronic eyepiece" which fits into a telescope drawtube. At $70 it is hardly a super-sensitive device like the cameras discussed above, but it does allow the Moon, planets, and even daytime objects to be viewed on a TV monitor.

Contact details for all these cameras are in the Appendix.

Sensitivity and Signal-to-Noise

When you consider that these cameras are working at 50 frames per second, interlaced to produce a 25 frame per second video, it may seem staggering that the exposures are long enough to get this deep into the night sky. But, ultimately, everything boils down to the signal-to-noise ratio.

Now let's have a bit of fun (!) with photon counting. The number of photons passing through a one-square-centimeter aperture, across the spectral range of a typical astronomer's CCD, is about one million per second from a zero-magnitude star (assuming typical atmospheric transparency/telescope optics).

This figure is no more than a good rule of thumb made to illustrate a few salient points! As 5 magnitudes = 100×, a 5th mag star will be responsible for 10,000 photons per second, per square centimeter. A 10th mag star will produce 100 photons per second, and a 15th mag star will produce one photon per second.

A 50-mm, f/1.8 lens (typical for a 35-mm SLR) with an aperture of 28 mm (6 cm^2) will actually collect 1500 photons per second from a 9th magnitude star. If we divide this by the effective exposure time of a video camera – let's say 1/60th of a second as an example – we get 25 photons per exposure arriving at our CCD. Far from being surprised that a video camera with this 50-mm lens gets down to mag 9, you might start to wonder why it doesn't go even deeper.

Indeed, when you consider what faint stars the human eye can see when you use binoculars, you realize that shortage of photons is not a problem: signal-to-noise ratio is the problem – but the human eye and brain is a superb signal processor! It's easy to get too fixated by the term "quantum efficiency" in relation to imagers and assume that if a CCD is 50 percent quantum efficient, it will happily display every second photon it receives, as a white dot on their computer screen. Dream on.

Although 25 photons might seem enough, in astronomy it is the signal-to-noise ratio that is the most important consideration – and the read-out noise and thermal noise of the CCD, along with the background sky brightness, are quite likely to swamp the signal resulting from converting this small number of photons into electrons. And how many pixels is your star image spread across? For long focal-length instruments this needs to be considered – the atmosphere alone will ensure that a star image is blurred over at least 10 square arc-seconds.

The calculations above show that capturing a mag 9 star is quite possible with a 50-mm lens and a video frame exposure. It's a sobering experience to attempt to record such magnitudes with a digital SLR using the same lens, where exposures of many seconds (or even minutes) are required.

Medium-Resolution Planetary Imaging

You have already seen that planetary imaging can be carried out with video cameras; more frames can be secured than with a standard astronomical camera, but the individual frames are noisier. Apart from this aspect the criteria for getting good planetary images are the same as discussed in Chapter 8. Go there for more information.

Total Solar Eclipses

Modern camcorders are ideally suited to capturing not only the detail but also the social atmosphere of a total solar eclipse. A single image can never do such a constantly changing spectacle any justice. Also, as anyone who has ever tried photographing a total solar eclipse will confirm, the usual experience is this: massive effort and panic during the eclipse; a feeling of remorse just after totality ends (because so little time was spent enjoying the eclipse visually); then finally disappointment when the prints are developed because they are all blurred. Or maybe that's just me.

Modern camcorders have an impressive dynamic range (ability to record bright and dim light levels on the same image) and most offer the facility to grab a frame and turn it into a single digital image, which can be imported to a PC. Although such an image will be noisier and have fewer pixels than an image taken with a quality digital camera, the camcorder is running all of the time and so will capture the moments of second and third contact (diamond ring) with ease, allowing the eclipse chaser to just sit back and enjoy the spectacle. The audio track is just as important as it will record the exclamations of your fellow travelers and reveal any surprises that occurred in their original immortal prose: "Wow, look at that prominence!", and "Look, aren't those shadow bands?", or "What a long diamond ring."

Analysis of the video can sometimes be used to extract some scientific data. If a real-time clock is displayed on the camcorder and the exact latitude and longitude are known (from a large-scale map or GPS receiver) the precise start and end of the eclipse can be timed and the duration deduced. This is especially useful if observers are at the extreme northern and southern limits of the totality band. At these points a lingering diamond ring, perhaps simply moving along the Moon's limb, will be recorded and the outer corona may never be fully seen. However, exactly what is seen is so sensitive to the solar diameter that analysis of the video tapes can actually help to see if the Sun is growing or shrinking. This type of measurement is hard to perform other than during a total solar eclipse; the Moon's orbit and profile are known precisely and it works as an occulting disk above the atmosphere.

The Sun's disk is so blindingly brilliant that even the most advanced optical equipment on Earth cannot easily measure the Sun's diameter to the precision that is achievable during a total solar eclipse. For many years, the International Occultation Timing Association has positioned amateurs along the north and south edges of total eclipse tracks and valuable science has been carried out. And the only tools needed are good-quality camcorders and GPS receivers!

Grazing Lunar Occultations

Every night the Moon drifts eastward among the stars by its own diameter (half a degree) every hour. In the course of this journey it, not surprisingly, passes in front of some bright stars. If the Moon's phase is earlier than full, the star will disappear at the dark limb and reappear at the bright limb. The opposite occurs after full moon.

Obviously the brightest stars along the Moon's path will be occulted least often because they are simply so few in number. If an observer is in the right place at the right time, a star may graze the Moon's mountainous north or south limb, i.e., the star will wink on and off as it disappears behind mountains and reappears in valleys! This can be quite dramatic but it is essential to be positioned on the graze track: to get the best effect one has to be positioned, in a north–south sense, within a few miles/kilometers or even a few hundred yards/meters of the position where the Moon's mountain's "shadows" will cause this effect. This will rarely happen just by sheer chance, at the observer's own site. So as with total solar eclipses, portable equipment is of great use, as is a sensitive security camera or camcorder to record the event. Domestic camcorders can be used, but, as we have seen, even more sensitive cameras are available at a modest price.

Organizing an expedition to observe and video grazing lunar occultations can be great fun. Of course, some research is inevitably necessary before setting out. The local farmer will not be happy if a dozen amateur astronomers suddenly string themselves out along his field, and the police may initially see a threat in the same number of people strung out along the roadside at night!

However, organizational (and weather) factors aside, a successful video of a grazing lunar occultation is a thing to be treasured. Few things in astronomy generate as many "ooh's" and "ahhh's" from an audience as a good graze video.

Planetary Occultations

Not surprisingly, the disappearance of a major planet behind the Moon is a rare event. In the UK, with our predominantly cloudy skies, seeing one can be a once-in-a lifetime event if you are really unlucky! Although the Moon drifts eastward through the constellations of the ecliptic, its orbit is tilted by 5 degrees. Not only that, but the orbital pole precesses too (completing one revolution in a little under 19 years). This means that the Moon does not track precisely along the ecliptic each month: it departs from it by up to 5 degrees. When this tilt is aligned with the tilt of the Earth's axis the Moon can range between +28.5 and –28.5 Dec. When it's aligned the other way (due to precession) it only ranges between +18.5 and –18.5 Dec.

As the major planets tend to keep much closer to the ecliptic plane, it's clear why the Moon so often misses passing in front of them. The good news is that when the Moon is on the ecliptic, at a point where a planet just happens to be too, it is quite usual to have a whole series of occultations during that year. As with grazing stellar occultations, grazing planetary occultations can be seen if you position yourself right on the track. As an example – to prove that it's possible – the grazing occultation of Jupiter (the track of which passed over northern England on 26 January 2002) was caught in a video frame by Ray Emery. It is shown in Figure 5.8a. The Saturn occultation, three months later, can be seen in Figure 5.8b.

To be honest there is little to gain scientifically by videoing these occultations – but as with their stellar counterparts, they are great for showing at astronomy meetings!

Lunar occultations come in four types, as I mentioned: disappearances at the dark or bright limb and reappearances at the dark or bright limb.

Modern astro-imagers need to consider carefully the different aspects of these configurations as the event will be over in a minute or two and forward planning can mean the difference between a brilliant success or dismal failure. For disappearances, you can, at least, see where the planet is before it disappears. But if the disappearance is at the dark limb (i.e., before full moon) you won't be able to

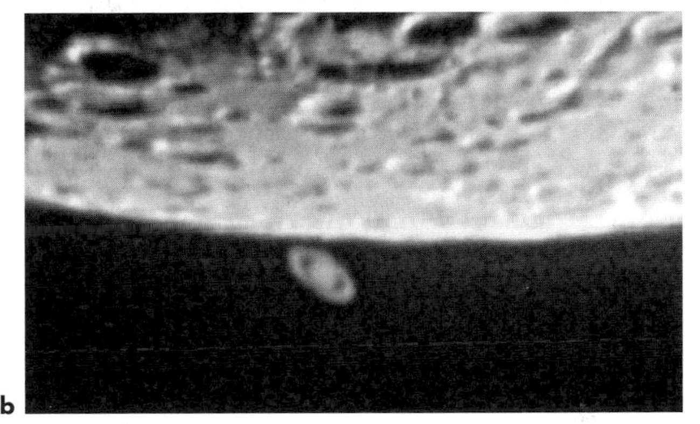

Figure 5.8. a The grazing north England Jupiter occultation of 26 January 2002. Image by Ray Emery using a Meade 250-mm LX200 with an f/6.3 focal reducer; Maplins security video camera, recorded onto Digital 8 tape.
b Saturn, just emerged from the lunar limb on 16 April 2002 at 21:26 UT. Image by Ray Emery using a Meade 250 mm LX200 with an f/6.3 focal reducer; Maplins security video camera, recorded on to Digital 8 tape.

see the limb approaching; this is fine if you are videotaping the event, as you won't miss a second, but photographers need to be on their guard. Even if the disappearance is at the bright limb, there may be problems. Take the case of Saturn and the bright limb of a full moon: you will probably want to have a picture with Saturn and its rings nicely exposed and a few lunar features well exposed too.

It may come as a shock when you realize that the surface brightness of the full moon is more than ten times than that of Saturn. Although Saturn is a much more reflective body than our dull, gray Moon, it is nearly ten times further away from the Sun, so it appears quite feeble next to the dazzling lunar limb. You will need to determine the longest exposure your system will tolerate without overexposing the lunar features, well before the night of the big event. It's amazing how easy it is to overlook things like this. With the lunar features well exposed, Saturn may appear very feeble in comparison, but remember that a bit of digital image processing can salvage an underexposed image, but not an overexposed one. For reappearances, the extra problem is knowing *where* the planet is going to reappear. A good astronomical handbook or planetarium software package will give the position angle (PA) of the reappearance on the lunar face for each astronomical event, e.g., a PA of 270 means the object reappears on the western (right-hand side with the naked eye) edge at the 3 o'clock position.

Once you have the PA you can look at what lunar seas or craters lie at this point and line your system up accordingly, well in advance. A CCD image of mine of the 16 April 2002 Saturn occultation reappearance is shown in Figure 5.9. No such luxury exists for a reappearance at the dark limb; all you can do is estimate where the planet is going to appear, based on the PA for your location and be on guard for when the first chink of the planet emerges.

This is one field where modern, sensitive security cameras or domestic camcorders are of huge advantage to the amateur astronomer. Being able to leave a camera running so that all you have to do is position the telescope is a major advantage. Digital camcorders nearly always allow single frames to be grabbed and saved in a TIFF or JPEG format for manipulation by image-processing software.

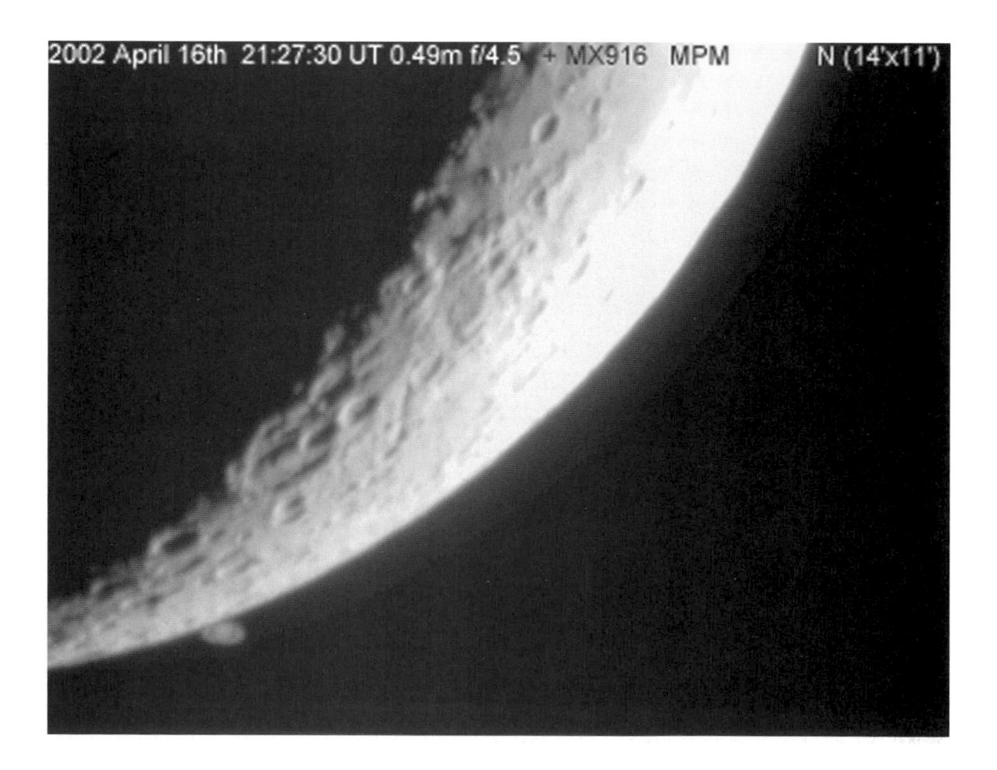

Figure 5.9. The same event as Figure 5.8b. Saturn emerging from behind the Moon. 16 April 2002 at 21:27:30 UT. Photo by the author with a 0.49-m, f/4.5 Newtonian stopped to 50-mm aperture(!) Exposure 1/100th second with a cooled MX916 CCD camera. Photo: Copyright Martin Mobberley.

Occultations of Bright Stars by Asteroids

In recent years the proliferation of powerful astronomical software and personal computers has enabled the easy and relatively accurate prediction of the occultation of bright stars (magnitude 12 and above) by asteroids.

When combined with bulk e-mail alerting of suitably equipped observers, this has led to valuable scientific data being collected, for example on the diameters of asteroids. In effect, when an asteroid occultation occurs, the observer is under the shadow of the star cast by the asteroid. Depending on the angle that the shadow makes with the ground, the shadow track on the Earth may have a width the same, or maybe double or treble that of the asteroid itself. Where the asteroid is very large and/or the asteroid's orbit and the star's position are known to great accuracy, the prediction of an occultation occurring (cloud permitting) can be made with some certainty. However, in most cases, the asteroid will be small (say, < 100 km) and the asteroid's position may be known to an accuracy of only one arc-second or so; in these cases the shadow track may be uncertain on a scale of as much as 1000 km.

Figure 5.10a shows an IOTA (International Occultation Timing Association) occultation track, in this case for the asteroid 345 Tercedina over Europe. The

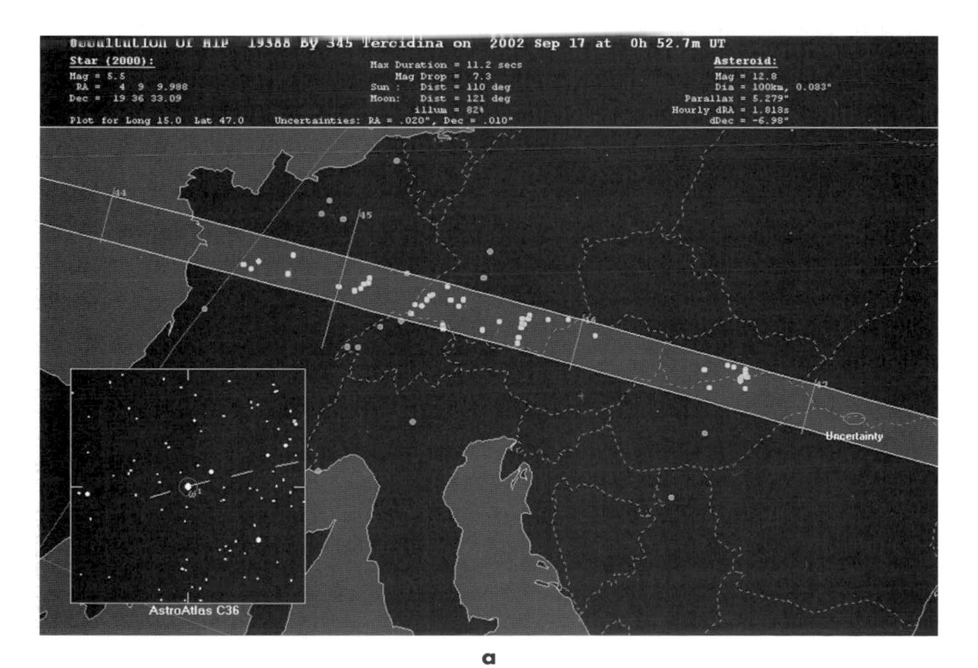

a

Figure 5.10. a The European ground track of the shadow of asteroid 345 Tercedina on 17 September 2002 as it occulted the mag 5.5 star Hipparchos 19388.The circles indicate the position of contributing observers. Diagram: courtesy Jan Manek/IOTA-ES.

(Figure 5.10. b, see overleaf)

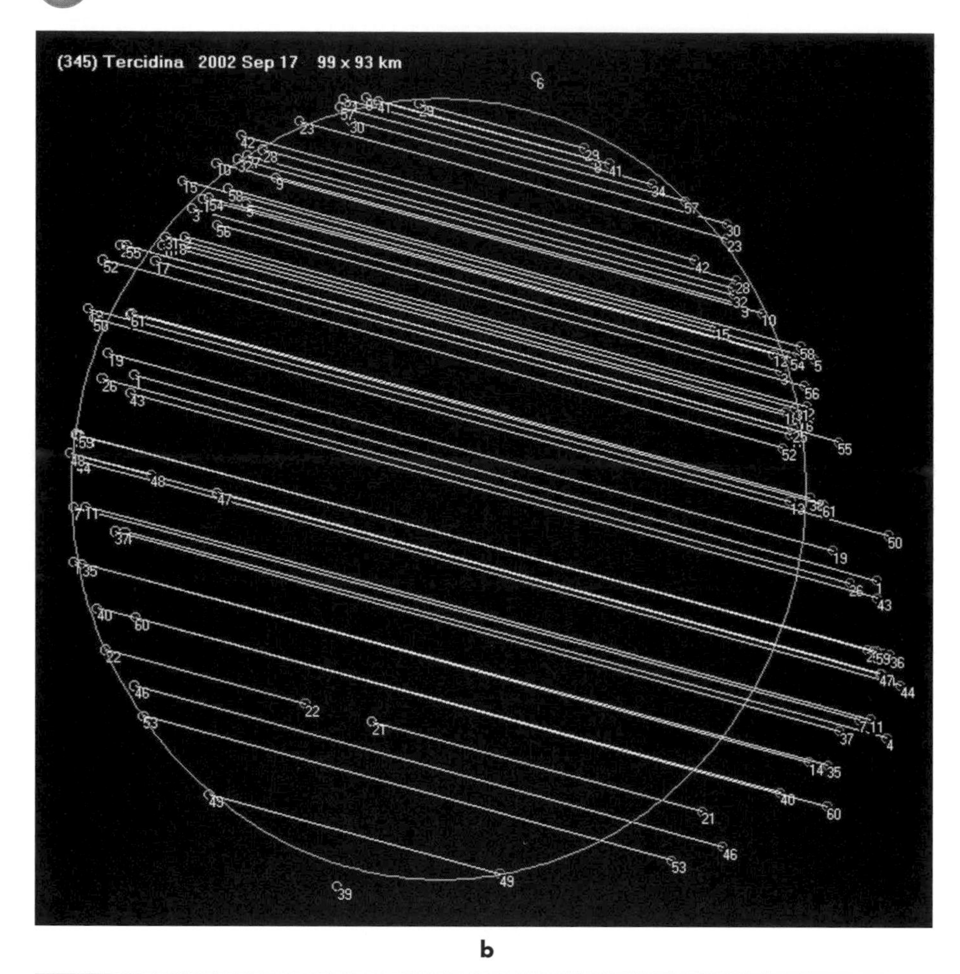

Figure 5.10. b The chords derived by European observers who timed the disappearance of the mag 5.5 star Hipparchos 19388 as it was occulted by asteroid 345 Tercedina. A total of 57 observer's chords pinned the asteroids diameter to approximately 99 × 93km. Diagram: courtesy Jan Manek/IOTA-ES.

observed disappearance chords are shown in Figure 5.10b; this was the most successfully observed European asteroid occultation ever, and second only to the occultation of the star 1 Vulpeculae by (2) Pallas on 28 May 1983 which was observed from 130 stations across the southern USA and north-western Mexico. Dedicated asteroid occultation observers may observe a dozen or more potential events without seeing a star wink out, but a success like the Tercedina event makes it all worthwhile for them!

Even in the cloudy UK, asteroid occultations have been observed by a few dedicated observers. Video technology is of great assistance to the new amateur astronomer who wants to participate. Staring constantly at a mag 11 star for ten minutes, trying not to blink and resisting the temptation to look at your watch, is a tedious business. Plus, when the event occurs, how accurately can you hit that stopwatch at the start and end of the star disappearing? How much easier things are if you can video the event and have a radio time signal superimposed on the tape. IOTA and EAON (European Asteroid Occultation Network) (see the Appendix) can advise on equipment here. The sensitivity of the video camera is all-important. I have already mentioned Andrew Elliott's endorsement of the Watec 902H low-light camera which can record stars to mag 12 with a 25-cm aperture; this camera and its rivals are ideal for asteroid occultation work. Domestic camcorders, even those which proudly boast a "one lux" capability, can be far less impressive and may only record stars of mag 5 with a similar aperture at normal video exposures.

Another piece of apparatus worth thinking about is the Santa Barbara Instruments Group (SBIG) STV (see Figure 5.11). The STV is a unique piece of equipment which can act as an autoguider for long exposures with a separate CCD camera (or with film) and can also be used as an imager in its own right. A personal computer is *not* required with the STV, it's a stand-alone instrument. This, in itself, is an attraction to many, whose sole experience of CCD imaging is fiddling about in the damp and cold with a tangle of wires between CCD camera and computer.

It's worth mentioning safety here. With a typical amateur telescope/imaging set up, safety is a vital consideration that is sometimes ignored. It is easy to make an unintentional DIY electric chair where mains wires and dew-damp seating is involved. If it's essential to use mains power, an ELCB (Earth Leakage Circuit Breaker) provides a good measure of safety – you can buy them at hardware stores – and should be considered essential. There is no point in spending vast amounts of money on equipment and then being too dead to use it because you were too mean to spend $15 on an ELCB.

Back to the camera. Images from the STV can be stored on video tape, or in the STV's own onboard memory or on a PC if preferred. Download to a PC can also

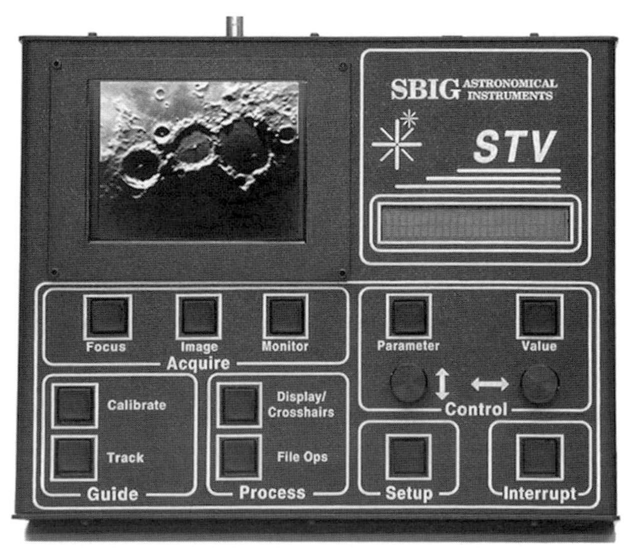

Figure 5.11. SBIG's versatile STV imager/autoguider. Photo by the author.

be carried out later, negating the need for a computer in the observatory. Essentially, the STV is a stand-alone, sensitive, cooled video camera, astronomical CCD camera, and autoguider in one. In the deluxe model a 5-inch (125-mm) LCD monitor is included so there is no need even for a separate monitor screen.

If you are thinking seriously about video recording and imaging asteroid occultations and other dim events, but would like to be able to instantly switch from video to long-exposure CCD work, the STV may be for you. It's certainly the only video camera I know of with a cooled CCD chip and the ability to increase the video exposure from 1/1000th to 600 seconds and store the result on video tape, internal memory, or a PC. Sensitivity is advertised as mag 14 in 1 second with a 20-cm aperture, which equates to mag 11 in 1/40th of a second; very similar to the video performance of one of the ultra-low-light security cameras mentioned above.

In passing, it is worth mentioning that a few wealthy amateurs use STVs attached to short-focus refractors, as the ultimate finderscopes when slewing their telescopes around the sky. SBIG's custom focal reducer "efinder" lens can be used as a finder in this context. It gives a 2.7-degree field down to mag 9.5 in a 3-second exposure. This, combined with the STV's quick download time, provides almost image intensifier-like finderscope performance, not dissimilar to using Adirondack's StellaCam-Ex attached to a camera lens.

Of course, many asteroid occultation observers will prefer to observe visually; it's certainly cheaper and you get the excitement of "I saw it with my own eyes" too.

I well remember seeing the occultation of the star 28 Sgr by Saturn's moon Titan on 3 July 1989. I observed the event visually through a 12-cm refractor while I videoed it with a primitive, experimental, low-light CCD camera I "borrowed" from work. This camera was valued at about $6000 dollars in 1985, but is out-performed by a typical $100 security camera in 2003!

Observing visually and adding an audio commentary to a video recording gives you the best of both worlds. In the UK, only a dozen or so observers have witnessed an asteroid occultation and the majority of successful events have involved just one observer. In a country where the skies are cloudy 80 percent of the time, and where there is, maybe, a 10 percent chance of the narrow track passing over the observer's location, this is hardly surprising. The bitterly cold night-time temperatures for half the year do not help generate enthusiasm either. Of course, from a scientific perspective, what is required is a series of "chords" across the asteroid's shadow profile, by observers situated along the (typically) north–south width of the asteroid track. The timings of how long the star was occulted from a specific location equate to a measurement of the asteroid's diameter at a certain "latitude" (for want of a better term) on the asteroid. If a whole set of accurate timings are available at asteroid "latitudes" from the top to the bottom of the asteroid, you have a set of chords which tell you (after a bit of maths) the precise shape of the asteroid. An example of this was shown earlier and an artistic rendering of what we are measuring is depicted in Figure 5.12.

There are only two other ways of measuring an asteroid's size and shape: send a space probe to it or bounce radar off one that comes within a few million kilometers! So asteroid occultations are a remarkably simple way that amateurs, with the minimum of equipment, can contribute real scientific results.

Okay, so how do you find out when the next asteroid occultation is occurring in your part of the world? In Europe, Ludek Vasta of the Czech Republic has set

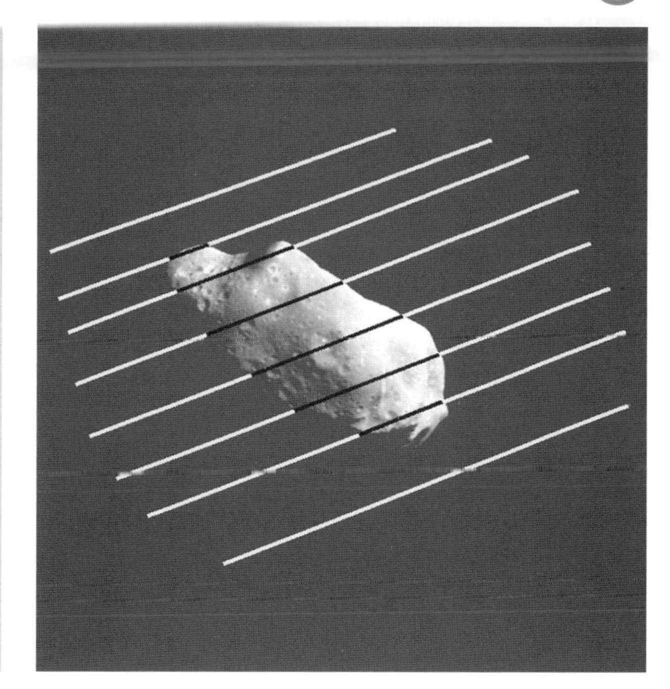

Figure 5.12. Measuring an asteroid's diameter by timing it occulting a bright star. Each black chord represents the period of time for which the star disappeared behind the asteroid, as timed by a stopwatch. Each line represents an observer at a location, say, a few miles north or south of the next observer. With enough observers, and accurate timings, an accurate profile of the asteroid can be deduced. Original image of asteroid 243 Ida: courtesy NASA.

up an excellent EAON (European Asteroid Occultation Network) Web site. The list of forthcoming events was completed by Edwin Goffin and was supplemented with the recommended observing window by Jan Manek. Internet addresses for these sites and a host of other useful Web sites is contained in the Appendix. You will also find a site for the world center for asteroid occultations, IOTA, run by Dr David Dunham, in the Appendix.

Videoing Meteor Showers

Astronomy is at its most dramatic when things happen quickly. The topics already covered in this section have dealt with events that change in a time-scale of minutes, but during meteor showers each bright meteor lasts only seconds, and may appear almost anywhere within the observer's field of view.

Unlike occultations and eclipses, the appearance of a really spectacular fireball is impossible to predict, even if one or two are almost guaranteed during the peak hours of the major annual showers.

The human eye, once dark-adapted, is incredibly good at finding and seeing both bright and faint meteors – it still rivals every form of man-made detector. For its aperture, the dark-adapted naked eye is incredibly sensitive; its only drawback is that it cannot integrate – long exposures are impossible. Experienced meteor photographers will be well aware this.

Even with a 50-mm, f/1.8 lens at full aperture, the fastest film will only record meteors of magnitude +1 or +2 at best – this despite the lens having an aperture advantage of three magnitudes over the pupil of the eye! Once again, we are back to quantum efficiency, which, in the case of photographic film, is only 2 percent at best, even without taking into account the noise from the background skyglow. Remember our calculation of photon flux in the section on grazing lunar occultations? Well, the human eye spans a spectral range of maybe half that of a CCD (for argument's sake) so the number of photons passing from a mag 0 star through a square centimeter of aperture at the Earth's surface is around 500,000 per second. An experienced observer at an exceptionally dark site can see stars as faint as magnitude 7.5 (and sometimes even fainter for eagle-eyed observers). This is hard to believe, but it is true. I struggle to see fainter than mag 5.5 with the naked eye from rural sites in the UK, but while these skies may look dark to a city-dweller they are not! Go to the middle of a desert or a few thousand meters up a mountain, well away from the nearest city, and you will see a completely different night sky – even the least experienced observer will see stars of magnitude 6.5 with no trouble at all.

If we take the dark-adapted pupil as having a diameter of 6 mm and we assume mag 7.5 is roughly 1000 times fainter than mag 0, this equates to 140 photons per second hitting the observer's retina. Because the eye's effective "exposure time" is only a fraction of a second and the brain will not register a "hit" unless several of the ultra-light-sensitive rods (cells in the retina) are struck, we can see that the human eye is a remarkable detector. In passing it is worth mentioning that some amateurs claim the eye does have an integration time of many seconds as, by staring for many minutes at objects just on the verge of their theoretical limit, they occasionally "glimpse" what they are looking for (presumably random variations in the photon flux suddenly allow them to see the object, or else their imagination comes into play!) Personally, that's not how my eyes work. I like to be able to hold an object steady with averted vision before I claim I can see it.

But I have digressed, as all I am trying to demonstrate is that contrary to what you might think, photographic film is a very *poor* detector of photons and the eye is a very good one. You need to hurl thousands, or tens of thousands of photons at a tiny point on your photographic film to leave an image.

Where meteors are concerned, long exposures don't help at all. They just fog the film with light pollution. Meteors streak across the film at an incredible rate, reducing the number of photons landing on any given point to a level, which – for most meteors – will leave no trace. Of course, if you can travel abroad to witness a meteor storm and if you use loads of cameras, you can get spectacular meteor photographs (see Figure 5.13), but for the average meteor shower we need to turn again to electronics to find a detector which is as sensitive as the human eye.

Image Intensifiers

Traditionally, an image intensifier was the instrument of choice for videotaping meteor showers. In this type of device, an incoming photon is converted into an

Figure 5.13. A composite of two images of a Leonid fireball, photographed from the roof of the Palasia Hotel, Palau, by the author at 18:48:30 UT on 18 November 2001. 50-mm, f/1.8 and 16-mm, f/2.8 lenses; Fuji Superia 1600. Photo: copyright Martin Mobberley.

electron, amplified by as much as 100,000 times and then hurled at a phosphor screen, which glows as the barrage of electrons hits it. The only record of the event is the persistence of the phosphor screen so, typically, a video camera is used to stare at the screen and record the events.

In more expensive commercial systems the video camera CCD is interfaced directly to the intensifier screen by a fiber-optic bundle, which is much more efficient in transferring the light.

Cheap intensifiers and ex-military models can be acquired for a few hundred dollars but some caution is necessary in buying. The cheapest commercial intensifiers frequently have low gain or noisy images and are *not* suitable for astronomy; even worse, many low-light "intensifier" systems feature an infrared

LED to illuminate the night-time scene at a wavelength too far in the red for the human eye to pick up. This is fine if you are using an intensifier for studying wildlife at night, but not a lot of good for illuminating meteors in the Earth's upper atmosphere! Good quality second, or preferably third, generation image intensifiers can be usefully employed in astronomy, even if the images may appear rather noisy (in fact, the noise is hardly surprising when one considers the amount of gain involved!).

Interestingly, up to the last few years or so, CCD frames would have looked noisy in comparison to even a single frame from an intensified video image, when amplified to the same degree. So if CCDs are as quantum-efficient as image intensifiers, why can't they be used for imaging meteor showers? Up to a few years ago, the problem lay with the read-out noise of the CCD electronics. CCDs used in video applications typically refresh the image 25 times a second (50 read-outs per second for an interlaced image). The read-out noise for a bright television image is insignificant, but it is very significant when you want to boost the detail to the maximum in low-light work. When CCDs are used for long astronomical exposures the thermal noise and background sky brightness are the biggest sources of noise, but, for very short exposures read-out noise is the dominant factor. Image intensifiers don't suffer from this problem because the image is not "read out", it is simply displayed on a glowing phosphor screen.

In the last few years there have been major advances by manufacturers in reducing read-out noise, not least because of the explosion of interest in camcorders, digital cameras, and low-light security cameras. The upshot of all of this is that the cameras already mentioned in this section, i.e., the Watec 902H, the Supercircuits PC164C, and the CCTV 2006X, all have much lower read-out noise than their 1990s equivalents; indeed, they rival image intensifiers costing ten, twenty, or thirty times more. You might have bought a cheap security camera simply to video the image intensifier's screen, but today you could cut the intensifier's cost right out of the equation, and just use the security camera!

So what sort of results can you expect from either an image intensifier or a low-light security camera video of a meteor shower?

As with using CCD cameras, a compromise is necessary to trade-off sensitivity with field of view. Wide-angle lenses inevitably have short focal lengths and correspondingly small apertures; they will cover large parts of the sky, but will not record the fainter meteors. And as with digital cameras, an intensifier's photocathode detector and the security camera's CCD will be smaller than a frame of photographic film (*much* smaller in the case of a CCD): therefore, the field of view with a given lens will be smaller than that produced by a film camera.

Despite this limitation, the quantum efficiency of these detectors means that with a fast short-focus lens covering 30 or 40 degrees of the sky, and provided the lens aperture is 10 mm or more, they can detect meteors more efficiently than the human eye.

A large photocathode may have a diameter of 20–25 mm so, in this case, the imaging area is similar to a film-based system, but perhaps 100 times more sensitive! At the time of writing, an affordable (i.e., less than about $1000) ex-military second or third generation image intensifier will still outperform a

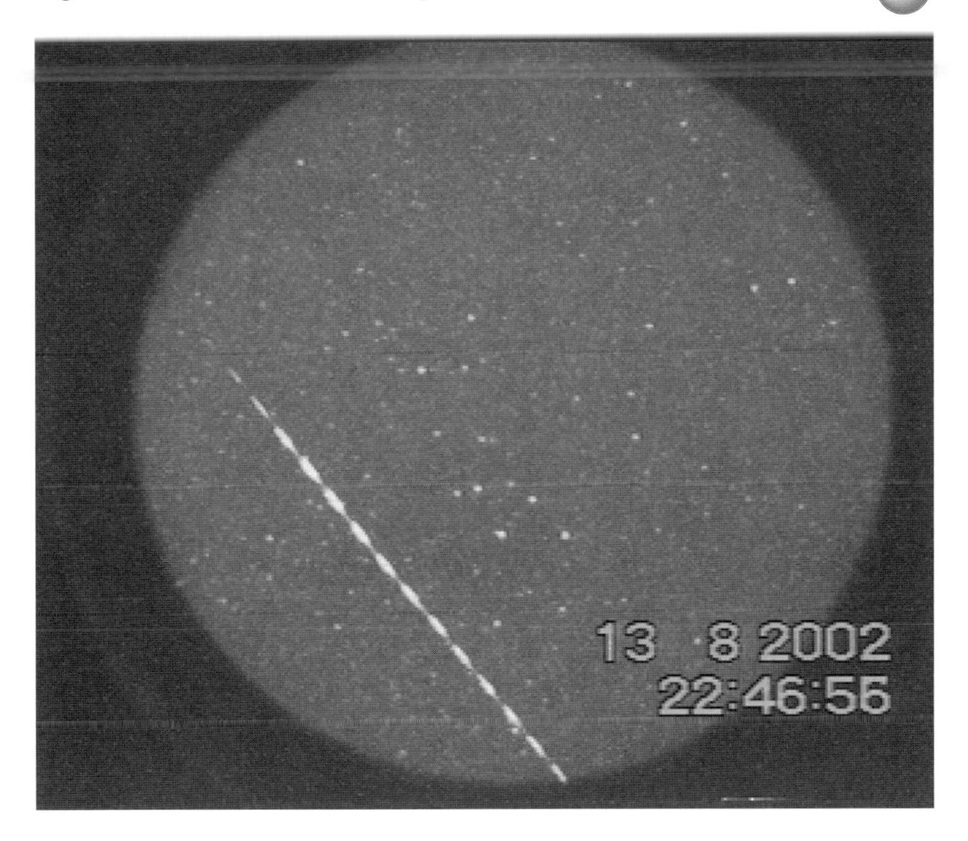

Figure 5.14. A typical image intensifier frame-grab. This is of a 2002 Perseid meteor. Photo: courtesy Steve Evans.

super-sensitive black and white security camera, but increasing improvements in the elimination of CCD read-out noise and increasing chip sizes will soon reduce the performance difference. A typical image-intensifier frame is shown in Figure 5.14.

In Germany, meteor enthusiast Sirko Molau has developed a system using a second-generation image intensifier and video camera, coupled to a PC equipped with a frame-grabber. His software, called *Metrec*, scans through frame after frame of meteor videos, automatically detecting and analyzing meteors! This software can be downloaded from the International Meteor Organization's Web site (see the Appendix). The *Metrec* system is as efficient as an experienced human observer at detecting and logging meteors – and it doesn't get cold and tired.

During the peak Leonid meteor storms of 1999 and 2001 some telescopic meteor observers even imaged faint flashes as Leonid meteors hit the lunar surface, using low-light video cameras. Sadly, by the time this book is published, the next Leonid storm will be some 30 years away.

Videoing the International Space Station (ISS)

Most amateurs will have spotted the Space Shuttle, the International Space Station (ISS), or the sadly departed *Mir* Space Station, moving rapidly across the night sky. Predictions for these events can easily be acquired from the Heavens-Above Web site (see the Appendix). A few amateurs have succeeded in imaging these objects at high resolution and resolving details as small as a few meters across. In terms of sheer resolution this is not a formidable challenge. For an object some 390 kilometers above the Earth, a simple calculation shows that 1 second of arc corresponds to just under 2 meters (tan (1/3600) × 390,000 = 1.89) at the distance of the ISS. Resolving details as small as an astronaut should be a formality for a planetary observer. Unfortunately, this is not the main problem.

The ISS is moving at about 8 kilometers per second to maintain its 90-minute orbit. This translates to a speed of 1.2 degrees per second (when it is overhead and closest), or some 4320 arc-seconds per second. Put another way, that's nearly 300 times the normal rate at which a telescope tracks the stars. Many commercial Schmidt–Cassegrains will happily slew at rates of up to 6 or 8 degrees a second, but these are not precise, programmable rates. So what techniques can you use to acquire high-resolution images of the Space Shuttle and the ISS?

Many amateurs who have succeeded in this area have invested in highly expensive telescope mountings such as the Software Bisque Paramount ME or the Merlin Controls Corporation Archimage mounting, or homemade mountings. Fortunately cruder and cheaper methods will work (with a little luck). A software package called *C-Sat* (see Appendix) is available which, running on a standard PC, will accurately slew a Go To Schmidt–Cassegrain to track any of 3500 satellites, including the ISS. Ron Dantowitz of the Hayden Planetarium, Boston, USA, has taken many fine ISS and Space Shuttle images with *C-Sat* slewing his 30-cm Meade LX200.

The one advantage that the amateur has in his favor in this field is the brightness of the ISS. It is a highly reflective object and, as it orbits the Earth, has no shortage of sunlight falling on it. The ISS is invisible in the middle of the night as it will be in the Earth's shadow when it flies overhead: satellites in low Earth orbits are only visible at dawn or dusk. (The Sun has to have set for the sky to be dark, but if the Sun is too far below the horizon, then no sunlight will fall on the satellite.)

Because the ISS is bright, it can be filmed with a standard video camera at standard video exposures, i.e., exposures of 20 ms and faster. Even if the telescope isn't tracking the ISS perfectly, the short exposure may freeze the motion. Moreover, with video it is possible to acquire thousands of video frames in only a few minutes of taping. These frames can be analyzed later by playing the video tape in slow motion and then frame-grabbing the best frames. Using this method, determined amateurs have obtained surprisingly good images of the ISS by manually slewing a telescope and keeping the ISS on the crosshairs of a well-aligned finder, while the video camera is filming at the main focus. It's a process that involves considerable concentration and gritting of the teeth for the few

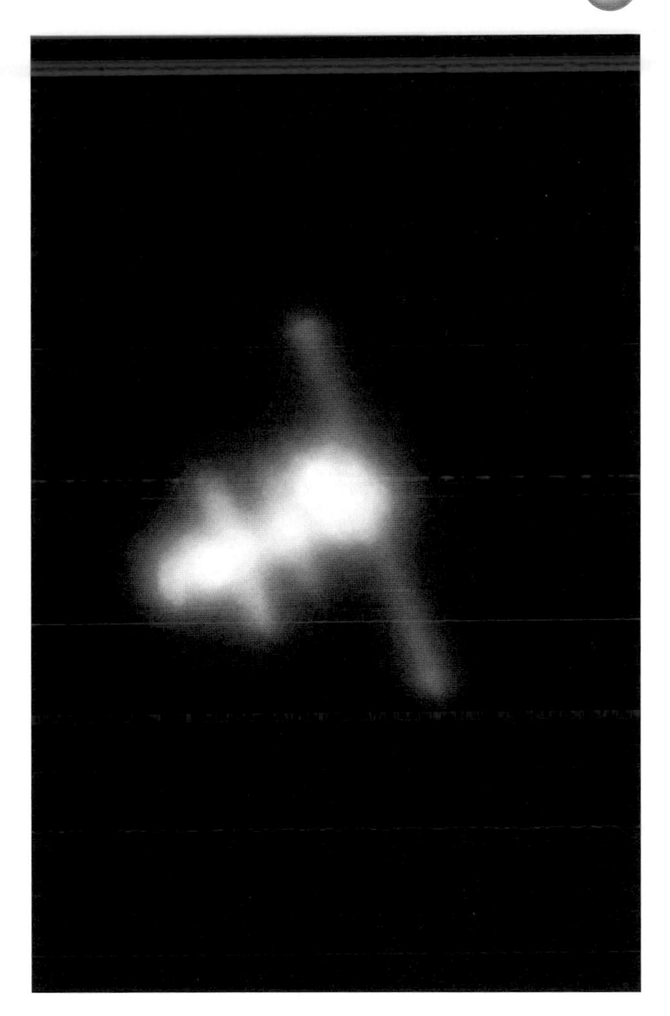

Figure 5.15. The International Space Station (ISS) imaged on 16 June 2002 by James W. Young and Gary L. Grasdalen. JPL/Table Mountain Observatory with a 24-inch (61-cm), f/16 Cassegrain and Minolta Dimage7 digital camera. A Scopetronix MaxView40 eyepiece plus EZPix digicam holder were employed to join camera to telescope. Exposure 1/15th sec at 100 ASA setting. The telescope was programmed to track the ISS.

minutes while the ISS flies overhead but, as so often in astronomy, there is often gain where there is pain. A good freeze-frame of the ISS with, perhaps, a Shuttle docking/undocking is a prized possession and another one of those shots that generates a lot of excitement at astronomy meetings.

A superb image of the ISS, taken with a Minolta Dimage7 Digicam, is shown in Figure 5.15.

CHAPTER SIX

Cooled CCD Cameras

Introduction

For the newcomer to astronomy scanning the magazines, it may well be more than a little confusing to work out what, exactly, a cooled CCD camera is. After all, digital cameras use CCD chips, too. Why are they so much cheaper?

Let us just review what a CCD does. Here's an analogy: a CCD is like a backyard full of empty buckets; several hundred rows of buckets times several hundred columns, making a total of at least a hundred thousand buckets (and often, more than a million) – okay, it's a big backyard. If we make a drop of rainwater the equivalent of a photon of light and call each bucket a "pixel," the analogy is complete. A CCD collects photons of light just like an array of buckets collects raindrops. The tricky bit is accurately measuring the water (the electrical charge resulting from each photon) in each bucket (pixel). This is an especially tricky problem in astronomy as the light levels are so low; it's like measuring a few raindrops in a damp old bucket, by pouring the contents through some leaky pipework! Digital cameras use the same CCD chip, but in normal daytime conditions the light levels are hundreds of thousands of times brighter – the buckets are almost always half-full.

When a small number of photons arrive at a CCD chip, they are almost swamped by the thermal noise in the electronics and, for short exposures, by the read-out noise of the CCD chip. Try taking a 30-second exposure of the night sky with a commercial digital camera and look at how noisy the image is – it's like a TV that's tuned to a station so far away that is almost out of range.

Fortunately, these problems are not insurmountable. For every 7 °C you can cool the CCD electronics, electrical noise is halved. Modern astronomical CCD cameras can cool the chip to 30 or 40 °C below the ambient temperature, resulting in a noise reduction of about thirty-fold.

This is not the full story. CCDs and electronics destined for use in astronomical cameras are specially manufactured to have the lowest possible electrical noise. They are also designed to have the highest possible quantum efficiency, ensuring that charge is generated in each pixel for almost every photon arriving. This technology comes, not surprisingly, at a high price. When coupled with the fact that individual astronomical CCD manufacturers only sell hundreds or thousands of cameras per year and Canon, Nikon, and Fuji sell millions of digital cameras per year, the price differential can be understood.

An attendant problem with cooling CCD cameras is what you do about the condensation. Astronomical CCD cameras need to be designed to be totally hermetic and, in many cases, have replaceable silica gel, to dry out the inside of the camera. We have seen elsewhere in this book that standard digital cameras can be used for lunar and planetary work (although webcams are even cheaper and better suited), but for deep-sky and comet imaging, even stacking hundreds of digicam images up cannot compete. The highest-quality CCD images of galaxies, nebulae, and comets will always be taken with cooled CCD cameras.

There are three different types of astronomical cooled CCD camera, namely:

1. Progressive scan CCDs (used in most high-quality astrocameras) which employ a mechanical shutter to cover the CCD chip when image download is taking place.
2. Frame-transfer CCDs (typically used in planetary imaging/small chip/autoguider cameras) which employ an initial fast (millisecond) image download into a light-shielded part of the chip, followed by a slower read-out from the chip to the measurement electronics.
3. Interline CCDs: low-cost chips, similar to those used in camcorders/digicams in which part of the pixel surface area is employed to pipe away the charge and the obscured area is partly reclaimed by microlenses on the chip surface. Starlight Xpress use this technology to the maximum, as we shall see later.

When affordable amateur CCD cameras first appeared on the market place, in 1989, the choice available to the amateur was very limited indeed. The first commercial success in this field was the Santa Barbara Instruments Group (SBIG) ST4. Initially intended as an autoguider for film-based, long-exposure photography, the ST4 soon became a popular imager in its own right, despite its tiny size (2.5 × 2.5 mm) and small number of pixels (32,000). Another small chip camera, the Lynx, also became popular around this time. Excellence in CCD imaging was not really a factor in the early years; it was just a novelty to be able to use a medium that was 10 times more sensitive than the fastest film and to be able to take exposures in 60 seconds and just trust the telescope drive to track for that period. Oh, the relief of not having to find a guide star and guide for 10 or 20 minutes (and don't even ask about offset-guiding for comet photography!) The downside of CCD imaging was simply centering the object on the chip, which often required nudging the telescope to an accuracy of a few arc-minutes! But now amateurs can have the best of both worlds: ultra-sensitive CCDs which are twice as sensitive as their predecessors and can have imaging areas almost as large as a piece of photographic film, if finances allow. In addition, the storage capacity of modern PCs and the speed of USB interfaces have allowed CCD technology to grow without encountering any serious technology bottlenecks. Parallel advances in Internet technology mean that amateurs around the world can upload their night's work to Web pages within minutes of the images being taken. It's certainly a refreshingly different astronomical world to that in the 1970s and 1980s where you often never saw the result of other amateurs' work until the photographs were published in magazines several months later! So what choices are available to the amateur CCD purchaser in the early years of the twenty-first century? Before we look at the technological goodies themselves we need to think about what we require from a CCD.

Choosing a CCD Camera

The most important aspect to consider when purchasing a CCD camera for astronomy is what field of view and what resolution are required. If your dream is to get the highest-quality deep-sky images then you will need to think in terms of one CCD pixel covering 1 or 2 arc-seconds (1/3600th of a degree) on the sky. In practice, in typical unstable air, 2 or 3 arc-seconds per pixel will suffice as even the faintest stars are bloated to 4 or 5 arc-seconds by the atmosphere and less-than-perfect telescope tracking. However, those who seek perfection in this field have demonstrated that with the best equipment, the best autoguiders, the best atmospheric conditions, and objects at a high altitude, image-scales of 1 arc-second per pixel can be justified, at the cost of field of view. Under these perfect conditions, faint star images may only be 2 arc-seconds in diameter! Unfortunately, the focal ratios of Newtonian telescopes are hard to adjust down to the optimum values, due to their f-ratios being quite fast anyway. However, accessories are available for the mass-produced Schmidt–Cassegrains working at f/10 or f/11 which telecompress their f-ratios by 0.63 or 0.33×, thereby increasing the number of arc-seconds per pixel and the available fields of view.

The simple formula below will save you time delving into trigonometry (remember a micron is a thousandth of a millimeter):

$$\text{Arc-seconds per CCD pixel} = 206 \times \text{pixel size in microns} / \text{Focal length in millimetres.}$$

For example, for a 490-mm aperture, f/4.5 Newtonian and a CCD camera with 23 micron pixels:

$$\begin{aligned}\text{Arc-seconds per CCD pixel} &= 206 \times 23 / (490 \times 4.5)\\ &= 2.15 \text{ arc-seconds per pixel.}\end{aligned}$$

If high-resolution planetary imaging is your goal, then you will need to think about much longer focal lengths, typically five or six times longer than for optimum deep-sky imaging. In this situation you will want to capture the finest detail that your telescope will ever resolve in the briefest moments of good seeing. The best planetary imagers on our planet have found that image scales of between 0.2 and 0.3 arc-seconds per pixel are optimum for capturing the finest and most exquisite detail (I will have more to say about this in a later chapter). This corresponds (due to the Nyquist theorem) to capturing details of 0.4–0.6 arc-seconds on the planet, or, put another way, the theoretical resolution of a telescope with a mirror between 290 and 190 mm aperture. If we choose an image scale of 0.3 arc-seconds per pixel, the following simple formula will tell us the focal length required:

$$\text{Focal length in millimeters required} = \text{Pixel size in microns} \times 688.$$

For example, for a CCD with 10-micron pixels:

$$\text{Focal length required} = 10 \times 688 = 6880 \text{ millimeters}.$$

For, say, a 300-mm aperture Schmidt–Cassegrain, this implies an f-ratio of 6880/300 = 22.9. Therefore, we could use a ×2 Barlow lens to increase the typical f/10 ratio of such an instrument to f/20, for optimum planetary imaging.

The same f/10 instrument and CCD camera could use a 0.33× telecompressor to give a 990-mm focal length and 2.08 arc-seconds per pixel, ideal for deep-sky imaging.

So a range of six times in focal length covers the requirements of both the high-resolution planetary imager and the "typical seeing" deep-sky imager.

Now we have the math sorted out, let's have a look at the toys available. Space does not permit a look at all the CCD cameras on the marketplace, but the market leaders' products are covered. We will start with the market leaders, SBIG.

Santa Barbara Instruments Group (SBIG) CCDs and Accessories

At the time of writing (late 2002) SBIG have a wide variety of CCD cameras available to suit most focal lengths and budgets. From the purchaser's viewpoint the most important considerations will be price, quantum efficiency, image download time, and matching telescope focal length to pixel size depending whether deep-sky, or planetary imaging, or both, is the aim. Another consideration is long-exposure guiding; there are three options here:

1. No long-exposure guiding, i.e., just short unguided exposures (the supernova patrollers' option).
2. Autoguiding using a separate CCD in the CCD camera head.
3. Adaptive optics guiding using SBIG's AO-7 device.
4. Autoguiding using a separate autoguider on a guidescope.
5. The old-fashioned way – manual guiding by eye (masochists only – horrendous).

Many of SBIG's CCD cameras feature a separate, smaller, CCD chip in the main camera head, specifically for the purpose of autoguiding. The movement of the chosen guide star across the guide-chip's pixels triggers correction commands to the telescope drive so that the star remains centered and the main CCD image looks sharp and without trails. However, a telescope has considerable mass and may take a fraction of a second to respond to the autoguider command. SBIG's

AO-7 adaptive optics unit uses the same information from the autoguider chip, but, in this case, the corrective data is not sent to the telescope drive, but to the mirror in the AO-7. It should be explained that the AO-7 unit fits between the telescope and the CCD camera and all the light heading for both CCDs has to go via the mirror. Because the lightweight mirror can respond instantly (within 10 ms) to commands from the autoguiding chip, the AO-7 will produce superb results with less-than-perfect telescope drives and will deliver better results at long focal lengths, even on good drives. The mirror in the AO-7 has a range of movement equivalent to about ±450 microns at the CCD or ±50 pixels with 9-micron pixels. If your image scale is 2 arc-seconds/pixel, this equates to ±100 arc-seconds or a total range of more than 3 arc-minutes. This means that if you have a poor telescope drive, but with less than 3 arc-minutes of total periodic error, you may well prefer to buy an autoguiding CCD camera from SBIG, along with an AO-7, rather than upgrade to a brand new telescope mounting.

SBIG's autoguiding CCD cameras currently come in five chip sizes, the ST-7XE, ST-8XE, ST-9XE, ST-10XE, and ST-2000XM. The "X" suffix stands for USB download; this feature was introduced to these five models in April 2002 and drastically improves the image download time. Three of these "X" cameras (the ST-8/9/10XE) now use a bigger autoguiding chip, too. Instead of using the Texas Instruments TC-211 chip as the tiny autoguider CCD, they now use a TC-237 CCD. This chip has three times the area coverage of the TC-211 and pixels that are twice as fine.

The ST-7XE is the lowest-priced autoguiding CCD camera from SBIG and features a 765 × 510 pixel CCD sensor with 9-micron pixels; this equates to a pixel count of 390,150 with an imaging area of 6.9 × 4.6 mm. Thus, from the formula above, the focal length required to achieve 1 arc-second/pixel is 206 × 9/ 1 = 1854 mm (73 inches) and for 2 arc-second/pixel, 927 mm (36.5 inches) is required. The field of view at 1 arc-second/pixel will be 765 arc-seconds × 510 arc-seconds = 12.8 arc-minutes × 8.5 arc-minutes, and at 2 arc-second/pixel it will be 25.6 arc-minutes × 17.0 arc-minutes. If your main astronomical interests and limitations are similar to mine, i.e., rapid, average-quality imaging, in poor seeing conditions (often through gaps in cloud) of topical novae, supernovae, and comets, you will find an image scale of 2 arc-second/pixel ideal for use with the ST7.

Why? Well, I have found in practice that in a country like the UK, astronomy is a battle against the elements and personal commitments. With less-than-perfect equipment and skies, and never enough spare time to do everything that a modern life demands, one needs equipment that is easy to use and will get to the target quickly and reliably every time, so any cloud gaps can be utilized. An image scale of 2 arc-second/pixel is more than adequate for such work and, with a field-of-view of 25.6 × 17.0, even a well-worn LX200 should get the target somewhere in the field!

The attraction of the original ST7 was the autoguiding chip, common to many of the cameras in the SBIG range (see the ST9XE autoguide system in Figure 6.1), built into the camera head; that's why I purchased one in 1997. On really clear nights, when no cloud is approaching, I can have a look at what the autoguiding CCD is seeing and pick a star for it to guide on. This process takes a bit longer than one is led to believe in the equipment reviews in the popular magazines, primarily because the aggressiveness of the autoguiding response needs to be tweaked to match the operator's system and the guide star brightness and, inevitably, there aren't always suitable guide stars in the autoguider chip's field (especially so at f/10!). Another factor here is declination backlash, which, on some nights, can make the whole process rather fraught! Of course, if one has the SBIG AO-7 unit, the telescope drives need not be troubled by the autoguider (unless the periodic error is so bad that the AO-7's mirror runs out of travel!). For those observers that specialize in getting half a dozen exceptional color astro-images per year, to 22nd magnitude, autoguiding with expensive drives and/or AO-7s is the *only* way to take astro-images.

I will have more to say about this breed of observers in Chapter 10. But for most of us and, especially the supernova patrollers, taking 60–120 second exposures and stacking them up is the simplest option on most nights. However, most amateurs will, I believe, want an autoguiding option because they dream of taking the perfect, long-exposure image. But, in practice, most will rarely use the option!

Figure 6.1. The main and autoguiding chip on the SBIG ST9XE, photographed by the author.

Figure 6.2. An SBIG ST8E image of the galaxy M101, by Gordon Rogers. Taken with a 16-inch (0.4-m) Meade LX200 at f/6.3. An SBIG adaptive optics AO7 autoguider was employed. Photo: courtesy Gordon Rogers.

With the new USB download, an entire 765 × 510 image can be captured in approximately one second, using a 1 GHz PC! So, if the SBIG ST-7XE is an excellent camera, why look at the more expensive SBIG autoguiding cameras?

The ST-10XE/10MXE and ST-8XE have loads of pixels (2184 × 1472 and 1530 × 1020, respectively), so an image produced by these cameras can have an almost photographic quality. A superb ST8 image of M101 by Gordon Rogers is shown in Figure 6.2. When combined with short-focus high-quality refractors or telephoto lenses, these CCDs can produce spectacular wide-field images of objects like the Andromeda Galaxy (M31) and the Orion nebula (M42). Such images will have so many pixels that they can easily be printed on the covers of astronomy magazines without looking pixellated or blurred. Although image scales of 1 or 2 arc-seconds capture the maximum information in an image, they often lack sharpness, because guiding and seeing problems make the smallest star images unpleasantly bloated. However, at a focal length of, say, 500 mm, the ST-10XE and ST-8XE will have image scales of 2.8 and 3.7 arc-second/pixel, respectively, so stars will look small, but not too blocky and the field of view (1.7 and 1.6 degrees wide) will be spectacular. The ST-10XE has very small pixels (6.8 micron) which makes it especially attractive to users of short focal-length instruments.

How about SBIG's ST-9XE CCD camera, pictured in Figures 6.3a and b? What's that all about? Big, 20-micron pixels and only 512 × 512 of them; why would you want to buy one of those? Here's why: the ST-9XE's large pixels are ideally suited

Figure 6.3. a The SBIG ST9XE CCD camera, photographed by the author.

for high-quality deep-sky imaging at the f/10 or f/11 focus of popular Schmidt–Cassegrain telescopes.

For the world's leading supernova patrollers, like Mark Armstrong, Michael Schwartz, Tim Pucket, and Tom Boles, the original system of choice was a 35-cm, f/11 Celestron 14 tube, mounted to a Paramount mounting with a big-pixel CCD. The Celestron 14 telescope has a focal length of about 3900 mm, so the 6.8- and 9-micron pixels of the ST-10XE/MXE and ST-8XE would give a third ($206 \times 6.8/3900$) and a half ($206 \times 9/3900$) of an arc-second per pixel at the f/11 focus, the sort of image-scale more associated with planetary imaging! Of course, the smaller pixels can be binned 2×2 or even 3×3 to make bigger pixels, but the critical point for many is that the big-pixel ST-9XE is half the cost of the ST-10XE/MXE and two-thirds the cost of the ST-8XE. The ST-9XE's 20-micron pixels give just over one arc-second per pixel at 3.9 meters focal length, an ideal image scale for taking advantage of good seeing on objects well above the horizon and seeking out those faint supernovae close to the cores of galaxies. If your main interest is in supernova patrolling, a camera with a fast image download time and an image scale of one arc-second per pixel is close to optimum. At one arc-second per pixel, the ST-9XE's 512×512 pixel array covers 8.5×8.5 arc-minutes – a small area of the sky, but just big enough for a Go-To LX200 to hit and an easy target for a quality mount like the Paramount or an Astrophysics GTO mount. A supernova image of mine with an ST-9XE at 1.4 arc-seconds per pixel is shown in Figure 6.4. An alternative option for taming the long focal lengths of Schmidt–

b

Figure 6.3. b The SBIG ST9XE CCD camera interfaces, photographed by the author, showing the power, USB, future accessories, and adaptive optics/color filter wheel/telescope guide ports.

Cassegrains is using a telecompressor such as Celestron or Meade's 0.63× focal reducer or Optec's WideField (0.5×)/MaxField (0.33× for small CCD chips) focal reducers. However, in practice, the most successful amateur supernova patrollers in recent years have not used telecompressors, but have opted for big-pixel, high quantum-efficiency CCD cameras; the results speak for themselves.

Another consideration for the CCD purchaser is whether to go for ABG or NABG (antiblooming gates or no antiblooming gates). What does this mean? An antiblooming gate is a semiconductor drain on the CCD surface which mops up the excess charge from an oversaturated pixel. In other words, when the light from a bright star fills up a pixel, the excess charge gets swallowed by the antiblooming gate (ABG). In a non-ABG (NABG) CCD the charge will overflow the pixel and, typically, leave a bright white vertical streak above and below the pixel. It might be thought that ABG was the best option; however, there is a disadvantage to ABG when the CCD is used for scientific purposes: the gates start draining excess charge well before the bucket is full; this means that the charge in the pixel is not related linearly to the brightness of the star if the star is more than 50 percent of the way towards saturation. So, if your prime interest is deep-sky pretty-picture imaging you may prefer a CCD with ABG, to avoid white lines bleeding from bright stars, but if you are interested in accurate photometry, you

Figure 6.4. A 60-second exposure of the Galaxy NGC 6946, with an SBIG ST9XE CCD camera and 30-cm LX200, showing the Supernova SN 2002hh. Photo: copyright Martin Mobberley.

may prefer the NABG option. In mid-2002, SBIG offered ABG and NABG versions on the ST-7XE and 8XE, but NABG only on the ST-9XE and 10XE. Another consideration is sheer quantum efficiency; ABG may get rid of ugly bleeding star images but it also reduces quantum efficiency. I would go for NABG every time.

Over the last five years or so, some of the best deep-sky images have been taken with the SBIG ST-7E and ST-8E with quality optics (often Ritchey–Chretien tube assemblies, but sometimes standard Schmidt–Cassegrains) and quality mountings, sometimes coupled to an SBIG AO-7 adaptive optics unit. Typically, the focal lengths of the telescopes have been tamed with 0.75× (Astrophysics) or 0.63× (Meade/Celestron) telecompressors to give an image scale of 0.7–0.9 arcseconds per pixel. But obtaining such quality is not just about equipment and image scale; it's about precise focusing, precise polar alignment, a lot of patience, and dark skies!

As yet, I haven't mentioned SBIG's newest camera, the ST-2000XM. This camera is a new departure for SBIG and their advertisements claim it was developed in response to customer requirements for a bigger and better imager, but with a lower price than the ST-8/10 models. The imaging chip used is an interline device (the sort used by competitors Starlight Xpress for many years). Interline CCDs are used heavily in video applications and so are much cheaper than their frame-transfer equivalents. SBIG have chosen to use a Kodak device; Starlight Xpress use Sony. Interline CCDs have circuitry to download image data *on the chip surface*, thus preventing all of the silicon being used to collect light. This used to affect the quantum efficiency substantially, but using microlenses to recover lost light on the chip surface means that these chips are now almost comparable in sensitivity with SBIG's smaller ABG Kodak chips, such as used in the ST-7XE. The 2000XM has 1600 × 1200 7.4-micron pixels, so it actually has more

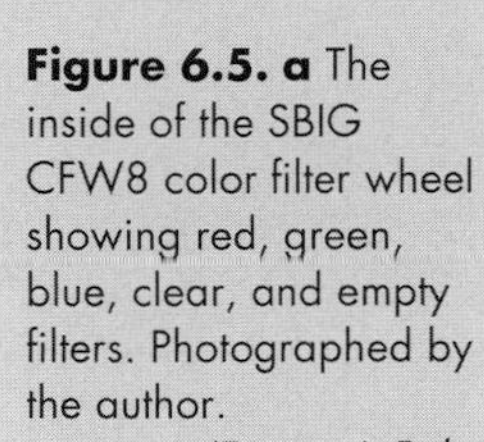

Figure 6.5. a The inside of the SBIG CFW8 color filter wheel showing red, green, blue, clear, and empty filters. Photographed by the author.
(Figure 6.5. b, see overleaf)

pixels than the ST-8XE, but a slightly smaller imaging area. However, the big point for many will be the cost: the ST-2000XM is only 60 percent of the price of the ST-8XE.

We have now had a good look at SBIG's autoguiding CCD cameras, all of which are compatible with SBIG's color filter wheel, the CFW-8 (Figures 6.5a and b) and the AO-7 adaptive optics unit. SBIG also makes four other cameras, the ST-1001E, ST-5C, ST-237A, and STV.

The ST-1001E boasts the largest imaging area of any SBIG CCD camera. It incorporates an active area of 1024 × 1024 24-micron pixels, i.e., 24.5 × 24.5 mm. So it covers 70 percent of the area of a frame of 36 × 24 mm film! It's a big chip with bix pixels, but is rather expensive and does not have USB download, compatibility with the CFW8 filter wheel, or an autoguiding chip.

The tiny pixel (10- and 7.4-micron) ST-5C and ST-237A come in a much smaller housing than the rest of the SBIG CCD cameras, making them less of a weight burden when used with small telescopes or piggy-backed telephoto lenses. Both the ST-5C and ST-237A can undertake deep-sky and planetary imaging, but the 320 × 240 pixel ST-5C has become renowned for its suitability for planetary work and the 657 × 495 ST-237A is particularly suited for use with the short-focus, f/2, Celestron Fastar system. Both these cameras are available with tiny internal color filter wheels but I fear their days as planetary imagers are numbered as inexpensive color webcams are taking over this area! However, if you have any doubts about the results that can be obtained with such small CCD cameras, just take a look at the tri-color images of Jupiter and Saturn obtained by Damian Peach in the section on planetary imaging! The fast download time of an image every 3 seconds used to make the ST5C the planetary CCD of choice, but

b

Figure 6.5. b The SBIG CFW8 color filter wheel attached to an SBIG CCD camera. Photograph by the author.

with webcams like the Philips ToUcam Pro downloading at 25 frames per second, this advantage has gone!

Developments in the CCD world and, especially, at SBIG rarely stand still for more than a few months. As the final proofs for this book were being checked, SBIG announced the imminent availability of some exciting new products for release in 2004. These included cheaper, non-autoguiding versions of their ST7, 9 and 10 cameras, a color version of the ST 2000XM, and a brand new line of Research Grade Cameras, featuring massive (up to 36 × 24 mm) chips, priced from $9,000 to $33,000! Check out the SBIG web page for the latest details.

Starlight Xpress CCD Cameras

In terms of value for money, Starlight Xpress CCD cameras have long been the pace-setters. This is especially so in the UK, where the company's founder, Terry Platt, has been a familiar face at astronomy meetings since the 1980s and his name is instantly associated with excellence in planetary imaging. Starlight

Xpress CCD cameras are small, lightweight, and affordable and offer a couple of one-shot color cameras that avoid the extra effort of tri-color imaging.

Affordability is the key selling-point of the Starlight Xpress cameras. They don't have cameras with a separate autoguiding CCD (although they have an autoguiding solution, called Star 2000) and they don't sell multi-mega-pixel cameras like SBIG's ST-10E. Their cameras are rather less sensitive than SBIG's latest models and some of their CCD chips have non-square pixels (which has to be fixed afterwards in software by resampling the image). Starlight Xpress use interline CCD chips, which are used in standard camcorders. In the more traditional astronomical progressive scan or frame-transfer CCD cameras, a larger proportion of the silicon is used for gathering light, because there is less interline circuitry on the silicon surface. At the time of writing, SBIG (see the section above) have also introduced an interline CCD camera (the ST-2000XM), although it is a lot more expensive than the smaller Starlight Xpress imagers.

Chip manufacturers have improved the light-gathering efficiency of interline CCDs in recent years, so there is less difference than there used to be. The Sony "SuperHAD" chips used by Starlight Xpress incorporate microlenses over each pixel, which direct light on to the light-sensitive area, that would otherwise fall on circuitry.

Another consideration, if you are thinking of using Starlight Xpress CCD cameras, is to understanding when a pixel is not a pixel! Chips used in video applications feature interlacing modes, where pixels may be "vertically binned" which halves the chip resolution. In addition, for one-shot color imaging, a 2 × 2 pixel color filter matrix is used to deduce the color allotted to each part of the image. Thus the luminance (monochrome brightness) resolution in these cameras is twice as good as the color resolution.

Because the human eye–brain combination is far more sensitive to luminance resolution than chrominance (color) resolution, it is a good, clean, high-resolution luminance component that makes an image look good; the resolution of the color component is less important.

If you were thinking of buying a Starlight Xpress color camera purely for short-exposure color lunar and planetary work, a webcam like the USB download Philips ToUcam Pro or the even faster Firewire download webcams may seem like a better bet. At prices as low as $50, these devices must be biting hard into CCD camera sales. Webcam-to-telescope adapters are widely available. But for cooled, long-exposure, one-shot color imaging, on galaxies, nebulae, and comets, Starlight Xpress still reign supreme, even if their color planetary supremacy has been "webcammed" away!

The main competition in all this is the Philips ToUcam Pro which uses a Sony ICX098AK interline CCD with 640 × 480 pixels. The chip's active area measures roughly 3.9 × 2.8 mm and has tiny 5.6-micron pixels. The colors are natural and download at rates as fast as a blistering 25 frames per second via the USB connection. The image on the PC looks just like what you see through the eyepiece. Small wonder that webcams are starting to dominate for short-exposure work.

Starlight Xpress cameras do not have silica gel compartments any more (the silica gel is used to dry the internal air to prevent condensation, when the Peltier cooler reduces the chip to –30 °C below ambient). Instead, the design utilizes a very small air compartment which seems to prevent moisture developing. Also, these small cameras do not include a mechanical shutter, so for taking dark frames you need to cap the telescope or the camera. With frame-transfer CCDs a mechanical shutter is often needed when imaging bright objects, such as the Moon, to prevent smearing during the read-out phase. Interline CCDs can do without them, thereby saving money and avoiding mechanical reliability issues.

At the time of writing, Starlight Xpress offer the following seven cameras:

1. MX516: 500 × 290 square pixels; 9.8 × 12.6 micron pixel size.
2. MX916: 376 × 290 square pixels; 23.2 × 22.4 micron pixel size.
3. HX516: 659 × 494 square pixels; 7.4 × 7.4 micron pixel size.
4. HX916: 1300 × 1030 square pixels; 6.7 × 6.7 micron pixel size.
5. MX5C: (color) 510 × 289 square pixels (luminance); 9.8 × 12.6 micron pixel size.
6. MX7C: (color) 752 × 580 square pixels (luminance); 8.6 × 8.3 micron pixel size.
7. MX716: 752 × 580 square pixels; 8.6 × 8.3 micron pixel size.
8. SXV-H9 and SXV-H9c: 1392 × 1040 pixels; 6.45 micron square pixel size. Option of monochrome or color.

Camera 8 was brand new as this book was going to press. The SXV-H9/H9c is, undoubtedly, Starlight Xpress' most exciting camera: a megapixel camera with the option of one-shot color and an optional, dedicated, inexpensive guiding camera. The Starlight Xpress web page will have the latest price and availability information. The SXV-H9 will probably eventually replace the HX916. Initial trials by Starlight Xpress show superb results from the one-shot color SXV-H9c.

As for the existing range:

Essentially, the range starts with a small and a big chip camera with non-square pixels. Then, there are two "HX" high-resolution cameras with small, square pixels and the two one-shot color cameras. Finally, we come to the latest camera, the mid-range MX716 which has small-ish slightly non-square pixels. All the cameras are contained in remarkably small barrels which are little bigger than a large "2-inch" eyepiece. The MX516 is the lowest price, entry-level model and, for many, the cheapest possible way to get into CCD imaging, apart from using a webcam.

The MX916 is a camera I have personal experience with over many years. Although the pixels are not quite square, resampling the image to correct the 3.6 percent difference in aspect ratio is not necessary; the pixels are near enough to being square that it makes no difference. The MX916 is often advertised as having a variety of high-resolution modes which double the resolution by using two interlaced frames in various ways. Some of these ways use interpolation to create another line, so the increase in resolution is rather artificial. Indeed, used at a focal length of 2.2 meters (2.1 arc-seconds per pixel) I have found that the high-resolution modes are indistinguishable from resampling the 376 × 290 pixels in Paint Shop Pro to give 752 × 580 pixels at an image scale of 1.05 arc-seconds per pixel! Resolution tricks apart, I have been very pleased with my MX916. The images are clean and noise-free and I often don't even bother taking a dark frame. Also, the M42 thread and lightweight nature of the barrel makes it very easy to attach camera lenses to the camera and easily mount the combination,

Figure 6.6. Supernova 2002ap and M74, imaged by the author on 14 February 2002 with a 0.49-m, f/4.5 Newtonian, from a light-polluted urban site, using a Starlight Xpress MX916 CCD and 180-second exposure.

piggyback, on a small equatorial mount. A 3-minute exposure of supernova 2002ap in M74, by the author, using an MX916, is shown in Figure 6.6. No dark frame or flat-field was used for this image and the only processing was a contrast stretch.

The software that comes with Starlight Xpress cameras allows the basic processing of images to be undertaken, but there are far better image-processing packages around, such as *Maxim DL*, *AstroArt*, and Richard Berry's *AIP*. Perhaps the biggest shortfall with the supplied software is its inability to display more than one image at a time and the refusal of the software to carry out all the functions on an enlarged image. Nevertheless, the software may be lacking advanced features but what there is, is very easy to use and instinctive to anyone with even a basic knowledge of CCD astronomy.

Another welcome aspect is that I have never had any problems installing Starlight Xpress software on any modern PC; unlike some other software, it doesn't need more than a few megabytes of hard disk space and most of the software will work with 8 megabytes of RAM. This may seem a trivial point, but many amateurs use their oldest PCs, laptops, or second-hand PCs, in the observatory and PCs more than five years old simply can't cope with the profligate memory requirements of a lot of modern software. Starlight Xpress software is extremely compact and so using it on your ancient observatory PC should not be a problem.

In many ways the MX916 embodies the advantages of the Starlight Xpress range: it's very compact, good value, totally reliable, and user-friendly. Plus, unlike with so many CCD camera companies, Starlight Xpress is not a faceless entity; if you e-mail them with a technical query, chances are that Terry Platt himself will e-mail you back in a few days with a solution. I know of no other major CCD manufacturer that offers such a service.

While I am on the subject of PCs it is worth mentioning that it saves an enormous amount of hassle if a PC can be kept in the observatory all the time. Even a laptop can be a nuisance to cart outside and connect up for each observing session, and power leads and data cables tend to mysteriously tie themselves into knots during the few hours they are outside. I've been using PCs and CCD cameras for over ten years now and can say, with feeling, that one of the most important criteria for successful imaging is hassle-free equipment. Numerous "obstacles" are placed in our way to prevent us taking excellent images: cloud, wind, dew, ice, trees, bushes, poor seeing, work commitments, family commitments, hard-disk crashes, and other equipment hassles are top of the list. To excel in imaging, these obstacles must be meticulously and systematically removed, one-by-one to reach the goal of excellence. Equipment hassles are often easily removed with a bit of thought. In the early 1990s I used to carry a full-size PC outdoors for every imaging session: it was utter human misery on a biblical scale! Then I wheeled a trolley with the PC on it outdoors, every night; this was still misery, but slightly less than would qualify it for the "utter and biblical" categories! Finally, I saw the light and extended the scope control wires and CCD camera cables indoors for both my 0.49-m Newtonian and 0.3-m LX200: utter bliss.

It is not always possible to remotely control a scope, and in these cases it is far better to keep an old PC outside in the observatory than to lug one outside each night. I know a lot of observers who use this strategy and the only pitfall is that the PC may get very damp when kept in the observatory, unless it is permanently switched on. While there is a tiny risk of any piece of electrical apparatus catching fire while it is switched on, there is an almost 100 percent certainty of temporary or permanent damage from cold, damp conditions. Keeping the PC on, permanently, will keep the damp at bay. The risk of the PC becoming defective and catching fire is not zero, though, and should be weighed against the advantages of it failing due to moisture ingress. Buy a smoke detector, and if it can be linked in to your home's security system so much the better.

With the sort of software that comes with Starlight Xpress cameras, the PC can be a *very* old one, so failure due to the damp, may not be a financial catastrophe.

Okay, now let's have a closer look at the square-pixel HX516 and HX916 cameras from Starlight Xpress. The HX516 CCD camera (Figure 6.7) incorporates tiny pixels and use advanced Sony "progressive scan" interline HyperHAD CCDs. With these chips there is no mystery about the resolution: the pixels are 7.4-micron squares and no resampling is necessary to make the image geometrically correct. The chip has a very low dark current and very good blue-light performance, making the camera excellent for tri-color work, useful if you don't like the rectangular pixels of Starlight Xpress' color cameras but want to achieve

Figure 6.7. The Starlight Xpress HX516. Photo: courtesy Terry Platt/Starlight Xpress.

an accurate color balance. For deep-sky imaging, the 7.4-micron pixels are particularly suited to focal lengths of around 1 meter (39.4 inches) where each pixel will span 1.5 arc-seconds. For planetary work, a focal length of 3 meters will give 0.5 arc-seconds per pixel; 6 meters will give 0.25 arc-seconds per pixel. UK imager Mike Brown has taken many superb lunar images with his HX516 and 37-cm Newtonian and one of his images is shown in Figure 6.8.

The HX916 CCD camera (Figure 6.9) has even smaller pixels and 1.3 million of them! Used in unbinned mode the camera has 1300 × 1030, 6.7-micron pixels; with the USB interface all 1.3 million pixels can be downloaded in 10–12 seconds. The size of the active area of the CCD chip is 8.71 × 6.9 mm, very similar to that of the MX916 CCD. In terms of number of pixels, the HX916 has only 20 percent less than SBIG's ST-8E, but the HX916 is a fraction of the cost. On the downside, the Starlight Xpress camera does not have a separate autoguiding CCD and it is not compatible with "Star 2000", Starlight Xpress' own autoguiding system. But like all Starlight Xpress cameras it is exceptional value for money.

Before covering the color cameras, I should just mention the Starlight Xpress MX716 (Figure 6.10). This camera sits in the middle of the Starlight Xpress range in terms of numbers of pixels (752 × 580) and in terms of pixel size (8.6 × 8.3 microns), and in these respects the camera competes directly with SBIG's ST-7XE camera, except that the SBIG model is a bit more sensitive, has precisely square pixels, and has an autoguiding CCD as well. Of course, the ST-7XE is also twice the price, which will be the critical issue for many. Although the MX716 does not have a separate autoguiding chip, it is compatible with the Starlight Xpress *Star 2000* self-guiding system. This system uses the interlaced structure of the Starlight Xpress interline CCDs to effectively use one interlaced field (i.e., every other row) for taking guiding images, while the other field stacks up light from the main exposure. Of course, one could argue that this is a cheat, as, in effect, only half the chip surface is gathering photons for the main exposure! However, Starlight Xpress have a point when they say that if you are autoguiding, doubling the exposure is not a problem! Nevertheless, the *Star 2000* system is a poor cousin to SBIG's dual CCD system. To use *Star 2000* you need to purchase a separate autoguider interface and the system only works with the MX range of monochrome CCD cameras (though MX color compatibility is promised). *Star 2000* has one big advantage and that is for guiding on a moving object like a comet. In any

Figure 6.8. A superb image of the lunar crater Theophilus and its surroundings by Mike Brown of York, England. Starlight Xpress HX516 camera with USB module attached to the motorized Crayford focuser of Mike's 37-cm, f/5.8 Newtonian. A 2× Barlow was employed. Photo: courtesy Mike Brown.

autoguider with a dual CCD, the guiding CCD sees a different star field to the main CCD. This is not normally a problem as all the stars are drifting by the same amount as the drive errors occur. However, if you want to track a moving object you need to track on that object and you can only guide and image the same object with *Star 2000*. So, dedicated comet imagers might be swayed by this advantage. Of course, if you are lucky enough to have a mount as good as a Paramount ME, you can let the perfect mount track precisely and program the system to track a moving comet without guiding at all; well, at least for a few minutes.

So what about the Starlight Xpress one-shot color cameras? The original images from the MX5-C and MX7-C with their chequerboard of color filters can look horrendous on the PC screen, but, after processing, things look much better. The color matrix pattern can make it very difficult to tell if you have a good or a bad image when you initially download it from the camera. You have to "synthesize" the color to get the full-color image. In this process, the signal strength data from each pixel under the color filters on the chip is analyzed and converted into a high-resolution luminance signal and a lower-resolution chrominance signal.

Figure 6.9. The Starlight Xpress HX916 Photo: courtesy Terry Platt/Starlight Xpress.

Figure 6.10. The Starlight Xpress MX 716 Photo: courtesy Terry Platt/Starlight Xpress.

Unfortunately the color balance needs tweaking, as does the color saturation, until the image looks natural.

Third-party software sometimes does a better job here and the excellent *Maxim DL* package has a superb command called "convert MX" which, when combined with the Maxim contrast stretch "planet" setting, produces instant quality planetary images (seeing permitting). The package *AstroArt* also copes with the MX color cameras. The big advantage with the one-shot color cameras is, of course, that only one image is needed, rather than three or four. In standard color imaging, amateur astronomers use L + RGB (L = visible white light, RGB = Red, Green, Blue) filtered exposures or, sometimes, IR + GB exposures, where IR is the unfiltered signal going to the CCD, with a strong infrared component. A stack of four one-shot color exposures of Saturn, taken with my own MX5c, is shown in Figure 6.11a. For deep-sky work these cameras offer a much simpler and cheaper method of taking color images of nebulae and galaxies (Figure 6.11b). A picture of the compact MX7-C is shown in Figure 6.12. If the color images taken with one-shot color cameras are compared with the best in the world, they may well look inferior, but compared to typical efforts by amateur astronomers using tri-color imaging they stand comparison. Amateur

Figure 6.11. a Saturn, imaged by the author, using the Starlight Xpress MX5c one-shot color camera on 28 September 2002. 0.36-m, f/20 Cassegrain and four 0.2-second exposures, stacked to reduce noise. **b** The peculiar galaxy M82 imaged by Terry Platt using a Starlight Xpress MX5c one-shot color camera on a 33-cm, f/4.3 Newtonian, with a 3-minute exposure. Photo: courtesy Terry Platt/Starlight Xpress.

Figure 6.12. The Starlight Xpress MX7c color camera. Photo: courtesy Terry Platt/ Starlight Xpress.

astronomers are often disappointed when they compare their efforts with the best images on the Web. But it should be borne in mind that the very best images are taken from the darkest sites, often at high altitude and with exceptionally long exposures, using autoguiding and/or adaptive optics units. While striving for excellence is fine, the very best results can rarely be equalled from a typical amateur's back garden. Certainly, in the UK, the vast majority of deep-sky images are taken of topical events, e.g., new supernovae or comets which may be at a low altitude, in light-polluted skies, and frequently in less-than-perfect sky transparency. Under such circumstances, perfect color balance will never be achieved and the difference between using a one-shot color camera and a tri-color system boils down to one factor: it's easier with a one-shot camera. All Starlight Xpress CCD cameras now have the option of a fast USB download and the color cameras are no exception.

We've now covered the cameras offered by the market leaders SBIG and the best value-for-money suppliers, Starlight Xpress; so what other cameras are out there?

Apogee Instruments CCD Cameras

Apogee is the company that brought the amateur community (at least those with the deepest wallets!) the first back-illuminated CCD cameras, in the late 1990s.

The company's AP7 CCD camera was more sensitive than any other camera available, even if it did cost more than a 30-cm LX200! The world's leading amateur supernova hunters were queuing up to buy this camera, when competitors' CCD cameras struggled to achieve half the quantum efficiency! An Apogee AP7, with camera lens attached, is shown in Figure 6.13. The AP7 also featured huge 24-micron pixels in a 512 × 512 pixel array; ideal for use at, typically, the f/11 focus of a Celestron 14, where each pixel would span 1.3 arc-seconds and the field covered was more than 10 × 10 arc-minutes. Nowadays, the situation is slightly different, as the best non-back-illuminated CCDs no longer look like "also-rans" compared with the AP7's quantum efficiency. So what, exactly, is this "back-illumination" process? Well, in a normal CCD chip, illuminated from the front, photons of light have to travel through an absorbing layer of semiconductor gate electronics before they drop into the bucket that makes up each pixel. Because of this journey, photons are lost, especially in the blue end of the spectrum. In a back-illuminated chip, photons come in from the other direction, avoiding absorption; but this can only be achieved if the chip is fabricated in an unbelievably thin wafer, about a hundredth of a millimeter in depth! Needless to say, making CCD detectors this thin is an expensive process, hence the high cost of back-illuminated CCD cameras. The chips used in the Apogee cameras are generally known as SITe chips, because they are made by Scientific Imaging Technologies. Apogee merely assemble the camera bodies and electronics – they don't even supply camera control and image-processing software. This is not a problem, though, as Software Bisque's *CCDSoft* and Cyanogen's *MaxIm DL*/CCD

Figure 6.13. The Apogee AP7 CCD camera body. Photo: courtesy Apogee Instruments Inc.

can be used for this purpose. There are other camera manufacturers who use these SITe chips in their cameras too; for example, in the UK, the Hale Research company use the SITe SIA502AB in their three stage water-cooled Hale EAC512-11 CCD camera, which is significantly cheaper than its imported Apogee AP7p equivalent.

Apogee also market a 1024 × 1024 pixel camera with a SITe CCD, namely the AP8p (p stands for parallel port). The larger camera also features 24-micron pixels and so covers four times the area of the AP7p. It is impossible to avoid comparison between these big-pixel cameras and market leader SBIG's own big-pixel ST-9XE and ST-1001E. The SITe chips are still the best in terms of quantum efficiency. SITe quantum efficiency typically peaks at almost 90 percent in the red and almost 70 percent in the blue end of the spectrum. In comparison, the Kodak KAF-0261E/1001E blue-enhanced chips peak at around 65 percent in the mid-visual spectrum, dropping to 45 percent in the blue. So the KAF chips are about two-thirds as quantum efficient as their SITe rivals. However, they are also less than two-thirds the cost of their rivals.

Apart from the AP7p and AP8p, Apogee also make non-back-illuminated AP cameras with the same Kodak chips used by SBIG and they now make cameras for medium focal-length scopes. The new AP47p has a 1024 × 1024 array of 13-micron pixels and uses a super quantum efficient, back-illuminated, Marconi CCD47-10 chip. At approximately 2.7 meters focal length (e.g., a 280-mm Celestron SCT at f/10), an image scale of 1 arc-second/pixel results. The AP10 (2048 × 2048 14 micron pixels) CCD camera is shown in Figure 6.14.

More details on all these cameras and many more can be found at the Apogee Web site (see Appendix).

Figure 6.14. The Apogee AP10 2048 × 2048 pixel CCD. Photo: courtesy Apogee Instruments Inc.

Finger Lakes Instrumentation (FLI)

FLI is a company that, in the last few years, has developed a loyal following amongst some of the world's best astro-imagers. The FLI camera bodies are compact (though not quite in the Starlight Xpress category!) and a number of cooling options are available. With noise doubling for every 7 °C increase in temperature, this is not a trivial issue and FLI have paid much attention to it; air, water, and turbofan-assisted cooling is available. FLI use the same Kodak CCD chips as SBIG for their front-illuminated systems (i.e., the KAF-0401E, KAF-1602E, KAF-0261E, and KAF-3200E chips), but they also offer back-illuminated Marconi and SITe systems including very large-format chips. At the time of writing there are two designs of FLI camera: the MaxCam series (Figure 6.15) and the IMG series (Figure 6.16). The MaxCam CM7, 8, 9, and 10 cameras use the four front-illuminated CCDs above and are therefore equivalent to the SBIG ST-7XE, 8XE, 9XE, and 10XE. The series also includes the MaxCam CM1 and CM2 cameras incorporating two back-illuminated chips from Marconi; a 512 × 512 matrix, 24-micron pixel chip and a 1024 × 1024 matrix, 13-micron pixel chip. The cameras are supplied with a 5-meter cable and FLIGrab camera control software. A separate "interface module" needs to be purchased too; these offer 16-bit parallel, USB, High Speed USB, or Ethernet communication with a PC. The high-speed USB download time is 3 microseconds per pixel, the same performance as SBIG's USB system. FLI can supply USB extenders for operating the MaxCam cameras at 100 meters distance.

FLI's IMG series offers superior cooling performance and parallel port operation over an impressive 150 meters without the need for signal repeaters or cumbersome external boxes to boost the signal. The IMG series also offer some mouth-watering large CCD arrays, for those with deep pockets! The IMG series also employ a guiding port, whereby another CCD can be attached (e.g., a MaxCam) and used as an autoguider for the main camera. The "dream machine" IMG512S and IMG1024S cameras include Scientific Imaging Technologies back-illuminated SITe SI512 and SITe S1003 CCDs. These are, not surprisingly, 512 × 512 and 1024 × 1024 pixel devices with 24-micron pixels and they compete head-on with Apogee

Figure 6.15. Finger Lakes Instruments MaxCam CCD camera. Photo: courtesy FLI.

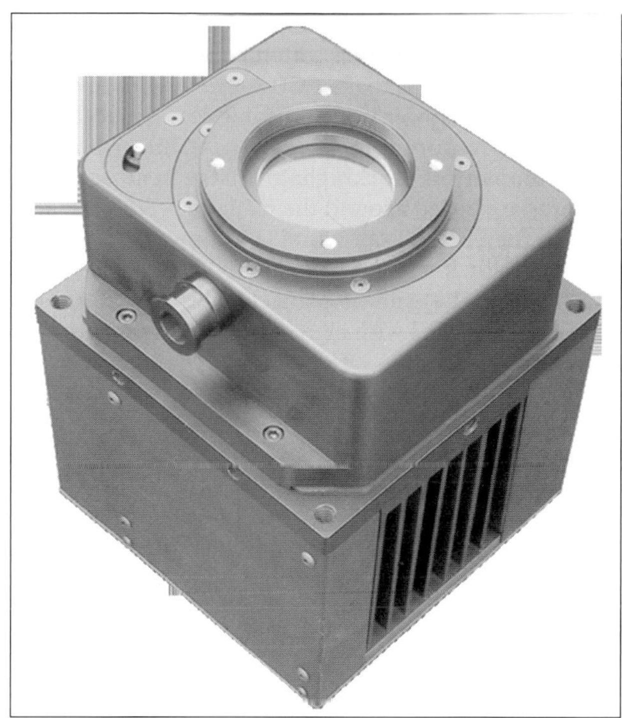

Figure 6.16. Finger Lakes Instruments Img CCD camera. Photo: courtesy FLI.

Instruments' AP7 and AP8 devices. At just under $7000 the 512 × 512 pixel device may well attract a fair few wealthy amateurs, but cameras based around the slightly less-sensitive KAF-0261E chip are much more affordable.

In passing, it's worth mentioning that FLI also offer a very nice digital focuser for use with their CCDs (Figure 6.17).

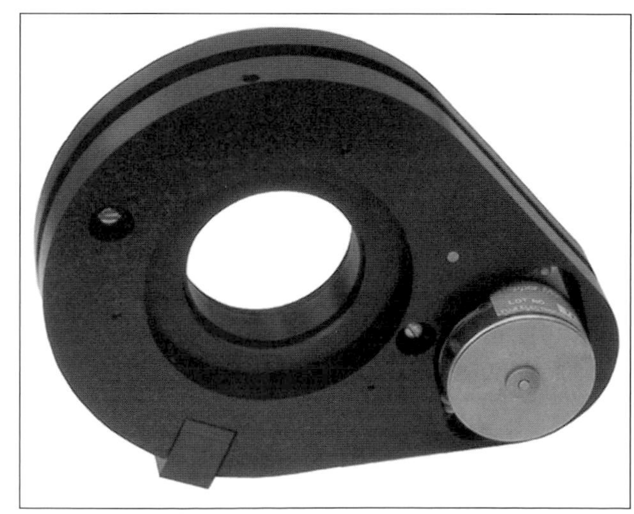

Figure 6.17. Finger Lakes Instruments Digital Focuser Photo: courtesy FLI.

Getting Through the Intimidating Image-Processing Jargon!

Unfortunately, software, and especially astronomy software, is loaded with abbreviations and jargon, much of which I will use in the coming pages. So to make things simpler, here is a brief explanation of some of the terms we will encounter in the manufacturers' blurb. If you want a full explanation I would strongly recommend purchasing Richard Berry and James Burnell's book *The Handbook of Astronomical Image Processing*.

After reading all 618 pages of that excellent book, your knowledge will be total! It is important to realize that you do not need to understand what all the jargon means. The vast majority of amateurs simply learn by asking other amateurs how they get their results and by "experimenting." Being a software genius is not necessary for processing images. Patience and the persistence to get the information out of the best amateur astronomers (who are always happy to help) is the best way to learn the art of astronomical image processing!

Astrometry

This is the technique of measuring star and comet positions by reference to known stars in the image. Advanced software can recognize the field stars and triangulate the position of the target with respect to the recognized stars. A catalogue (e.g., the *Hubble Guide Star Catalogue*) is usually provided with the software. Some software will even generate a minor-planet astrometry form for you. This will format your data in the style required by the Minor Planet Center, the central clearing house for all asteroid and comet astrometry.

Automated Dark-Frame and Flat-Fielding

Dark frames are exposures with the camera shutter closed. They are a map of the thermal noise of the CCD camera and are of the same duration as the image of the night sky. Subtracting the dark-frame results in most of the noise in the image being removed. A flat-field image is an image of a perfectly uniform sky background, e.g., a perfect twilight sky, just before the stars appear. Such an image will record all the deficiencies of the telescope (dust-specks and light lost due to vignetting). When the astronomical image (minus dark frame) is divided by the flat-field image, the resulting image looks perfectly smooth and clean. Many modern software routines offer the option of auto-flat-field and dark-frame operations; in other words, the camera can be programmed to routinely close the shutter and take a dark-frame and to call up a flat-field from memory. Every time a new picture is taken, the dark-frame and flat-field operations are routinely applied to save the user the hassle of applying them manually. Typically, a dark frame will be taken every time the camera temperature varies by a fraction of a degree.

Autofocusing

There are few things more full of hassle than battling with focusing a CCD camera. Tweak the focuser; check the monitor; tweak the focuser; check the monitor; it can be a real pain. Fortunately it is now possible to automate this using software such as "*@Focus*". In these systems the CCD image downloaded from the USB or parallel port is checked for sharpness (how small the star images are) and a motorized focuser (usually connected to the serial port) is controlled to minimize the star images. With USB downloading of a small part of the image, this can all be achieved in seconds, instead of minutes, or tens of minutes.

Blink Comparison

When searching for moving objects like faint asteroids it is useful to be able to blink two images taken minutes or hours apart. Some software offers this feature, and some will even align the images for you; the best one I've come across is in Richard Berry's *AIP* software. It is invaluable for if you are conducting a nova patrol.

FFTs and Kernels

These two terms are often seen in even the simplest processing manuals and can seem to be there to cause insecurity in beginners. An FFT is a fast Fourier transform. This is simply a quick way of representing any periodic mathematical function by an infinite (actually, a few dozen will do) sine or cosine functions. As all images have high- (small-scale) and low- (large-scale) frequency data, a way of representing and manipulating these data quickly is of great use: hence the FFT. FFTs are simply mathematical tools from which image-processing routines can be assembled.

The word "kernel" has no military connotation, it simply means "the nucleus, or essential part', as in the middle of a nut. In image-processing packages, such as *Maxim DL* and other products, the kernel filters are simply a set of high-pass (sharpening), low-pass (smoothing), and other assorted alternative filters to FFTs. The exact usage varies from package to package but kernel options often include filters for removing dead pixels (black) and hot pixels (white).

Filters

It is unfortunate that the term "filters" in astronomy software means something totally different from the colored glass filters used in the same hobby! In image processing, a filter is a confusing term for a mathematical operation carried out on an image. I will cover the DDP (Digital Development Processing) filter technique in greater detail at the start of Chapter 10. Essentially, it gives a CCD image a "film-like" quality, where the faint regions are enhanced, bright regions suppressed, and star images tightened, all with little increase in noise.

The unsharp mask filter is, undoubtedly, the most powerful routine used by the planetary imager, but it needs using with care so that artefacts are not created. The name dates from the era when photographers would create a blurred mask of an original negative in the darkroom. This mask would then be used as a real optical filter to suppress large-scale, low-frequency changes and enhance fine high-frequency detail. DDP has, in many ways, become the "ultimate" filter for deep-sky imagers, but there are others. They are all aimed at making the images, whether deep-sky or planetary, look sharper, but without excessive noise.

The process of "deconvolution" will often be mentioned in this context too. Without going into great detail and boring the reader into a coma, deconvolving is a bit like having a blurred image and trying to guess what the sharp original might have looked like. The software has a guess at what the original looked like and then blurs it to see if the blurred guess looks like your grotty image. This process goes through numerous iterations until the best guess emerges. In the 1980s this process took hours on the fastest home computers. Now, it takes seconds! There are various routines that are used: maximum entropy deconvolution is one, Lucy Richardson is another. Contrary to popular belief, Lucy Richardson is not a girl's name! This particular mathematical transform was simply invented by W.H. Richardson in 1972 and adapted for astronomy by L.B. Lucy in 1974.

A few software packages offer a "Larson–Sekanina" filter which is a mathematical transform specifically designed to show radial jets emerging from cometary nuclei.

LRGB

The LRGB technique, as applied to planetary and deep-sky imaging, is explained in detail in Chapters 8 and 10, respectively. The letters stand for Luminance, Red, Green, and Blue. A color image of any astronomical object can, of course, be created from the monochrome images taken through red, green, or blue filters. In effect, the software simply mixes the colors together. But in recent years, especially for deep-sky images, it has become popular to take a deep luminance (unfiltered light) image to get the maximum signal-to-noise and then simply color it using the R,G, and B data. The result combines the depth of the unfiltered image with the RGB color, but without the RGB noise. Many modern software packages allow hassle-free RGB or LRGB image aligning and merging. Some software offers CMY (Cyan, Magenta, Yellow) tri-color merging too. The CMY technique has gained popularity with CCD imagers in recent years because it offers better coverage of the visible spectrum and inputs more light to the CCD, but with RGB filter wheels being the norm it is a specialized technique.

Photometry/Light-Curve Generation

If you have the appropriate filters for your CCD camera, you may wish to measure the brightness of variable stars. Photometry software will convert the electrical charge collected in the pixels of your CCD into stellar magnitudes and allow you to specify an "aperture" around the stars in which the starlight and background light is measured. This is a precise science in itself! Recent software packages also allow automated photometry and light-curve generation. In other words, the software will measure the brightness of a target star, compare it to a known star, perform the math and repeat this throughout the night. At the end of an observing run the data will be presented in a form that can be loaded into a Microsoft Windows *Excel* spreadsheet and produce a graph of stellar magnitude versus time for the variable and for the comparison star.

Now let's have a look at some of the best software packages around.

Software Bisque's *CCDSoft, The Sky*, and *Orchestrate*

Software Bisque is the most respected name in commercial astrosoftware bar none. In 1984 the Bisque brothers, Steve and Tom, entered the astronomy arena with their ground-breaking software, *The Sky*, and have never looked back. The original version of *The Sky* was a simple planetarium program capable of displaying 1000 stars on a PC/Mac screen. Nineteen years later, the latest version allows access to over five hundred million astronomical objects. For the past ten years *The Sky* has been able to control telescopes like the LX200 by connecting the PC's serial port to the telescope. Software Bisque also produce a superb image-processing package called *CCDSoft* and, with their *Orchestrate* package, *The Sky* and *CCDSoft* can be orchestrated. What does this mean? Well, *Orchestrate* allows you to write an integrated script for the telescope and CCD camera so you can carry out, for example, an automated supernova patrol. Such a script would define the CCD exposure duration for each galaxy, the list of galaxies to slew to, and the directory in which to store the images. In theory you could go to bed while the telescope was patrolling! The UK supernova discoverers Mark Armstrong and Tom Boles use Software Bisque's integrated package to control their telescope mountings and CCD cameras in this way, but, in reality, they don't go to bed – they watch the images downloading as they sit indoors in the warm! Both observers can secure over one thousand images *per night* with this system, with each CCD image reaching fainter than mag 18 in a 30-second exposure! As described elsewhere in this book, Software Bisque decided to enter the world of telescope mountings in 1996; they already supplied software to automate telescope mountings, but there was no robotic mount out there that could do full justice to their software! So, in 1996 they produced the Paramount GT1100, later upgraded to the GT1100S and, in 2002, the superb Paramount ME. Whatever area of astronomy Software Bisque have entered they have demonstrated excellence.

CCDSoft Version 5 is a powerful image-processing package which comes supplied with all of SBIG's CCD cameras. (SBIG also supply their own CCDOPS software, but most users will prefer to use *CCDSoft*. SBIG also supply *The Sky* with their cameras as well, a total software bundle of nearly $500!) As well as being compatible with SBIG's CCD cameras, *CCDSoft* can also control Apogee and Finger Lakes cameras, as long as an additional software plug-in is purchased.

CCDSoft's features are simply too numerous to mention in their entirety, but they include:

- Comprehensive LRGB image combining/filter wheel control commands;
- The *@Focus* autofocusing software (with compatible focusers);
- All the usual contrast/brightness histogram features;
- A blink comparator plus astrometry/photometry/light-curve routines;
- Automated minor-planet astrometry form generating;
- Unsharp mask and Lucy–Richardson processing;
- Numerous filters and pixel operation processes;
- Batch processing for auto-flat-field and dark-frame operations;
- A useful quick-zoom feature plus movie creation!

While not actually an image-processing program, *The Sky* includes the following planetarium features:

- Displays the night sky from 4712 BC to AD 10000;
- Shows stars down to mag 16 (Hubble) Guide Star Catalogue (GSC) or even mag 20 (Palomar) Deep Sky Survey (DSS);
- Displays the moons of Jupiter and Saturn plus a 3D Solar System view;
- Produces star charts;
- Plots artificial-satellite tracks;
- Integrated telescope control and *CCDSoft/Orchestrate* compatibility.

Without a doubt, Software Bisque's trio of CCD and telescope software packages is the most powerful system around, especially for those with SBIG, FLI, or Apogee CCD cameras. But there are other packages around which may suit those with other cameras, or serve as an extra option for those who like to have every software tool at their disposal. Let's now have a look at Richard Berry's *AIP*.

AIP

With most things in life, you get what you pay for, and the best rarely comes cheap. One exception to this is Richard Berry and James Burnell'*AIP* (*Astronomical Image Processing*), which comes in the form of a superb book with a software CD attached. I am sure that most *AIP* readers will forgive me if I say that, for me, it is the image-processing CD-ROM that comes with the book that attracts most attention. But first, let's deal with the book itself.

Few amateur astronomers will not have heard of Richard Berry. Since the late 1980s he has encouraged amateurs to use and understand their CCD cameras, via his articles and books (*Choosing and Using a CCD camera*, *Introduction to Astronomical Image Processing*, and *The CCD Camera Cookbook*). His collaborator for the *AIP* book, James Burnell, became seriously interested in image processing in the late 1980s, when he was working for Bell Labs.

His first experiments with this technology were involved in playing with NASA's voyager images using a UNIX mainframe computer. After meeting Richard in the early 1990s, the two decided to collaborate on writing *The Handbook of Astronomical Image Processing*, a project that took three years to complete.

The book itself, at 624 pages in length, would be remarkable value as a book alone, even without the software. As an author myself, and being very familiar with the small profit margins of specialized astronomy books, I can say with confidence that Berry and Burnell have written this as a labor of love – a service to the astronomical community.

The book is divided into 18 main chapters which take the reader from concepts such as basic imaging, digital image formats, and how to make good CCD images, through to specialized fields such as astrometry, photometry, and color imaging. The four appendices contain a glossary, information resources section, software functions listing, and an excellent step-by-step tutorials section for the software on the CD.

There is no easy way to teach the fundamentals of image-processing techniques to an absolute beginner, but the writing style, diagrams, and explanatory figures in the book could not be better. If you have the time (how many of us do?) to read and digest the whole text you would, without doubt, understand all that there is to know about astronomical image processing. This is not a book that shies away from trying to explain concepts totally. If you really want to know how many photons are hitting each pixel on your CCD camera every second, how many electrons are produced, how much noise affects the image, and how image-processing math works, then it's all in this book. There really is no other book covering the subject of image processing in such depth and clarity.

So, moving on to the software itself, I suspect that most amateurs will buy the book for its "*AIP4Win*" CD. If the book were not good enough value in itself, how does a powerful image-processing package for $79.95 sound! If you look at the competition (e.g., *Maxim DL* at $300) the book/software package really is astounding value. As far as I can see, every image-processing routine you might need is in the package! I think most users of image-processing software will crave a number of features beyond all others; few CCD users have not at some point "lost the will to live" with frustrating software packages. I think that the key features we all look for in an image-processing package are:

1. An intuitive user interface;
2. The ability to import all image formats from all CCD cameras;
3. The ability to do some science with the images;
4. The incorporation of as many powerful routines as possible.

I am happy to report that *AIP4Win* satisfies on all four counts.

I've played with the *AIP* software for a couple of years now and it's definitely the best package that I've come across at this price. Space does not permit me to detail all of *AIP*'s features, but every useful image-processing routine I know of is supplied. When I first acquired *AIP*, I was initially a bit concerned that the relatively recent (in 2001) arrival of Kunihiko Okano's powerful Digital Development algorithms on the deep-sky scene might have prevented their inclusion on the CD. I need not have been worried – *AIP4Win* has a superb DDP tool (Figures 7.1

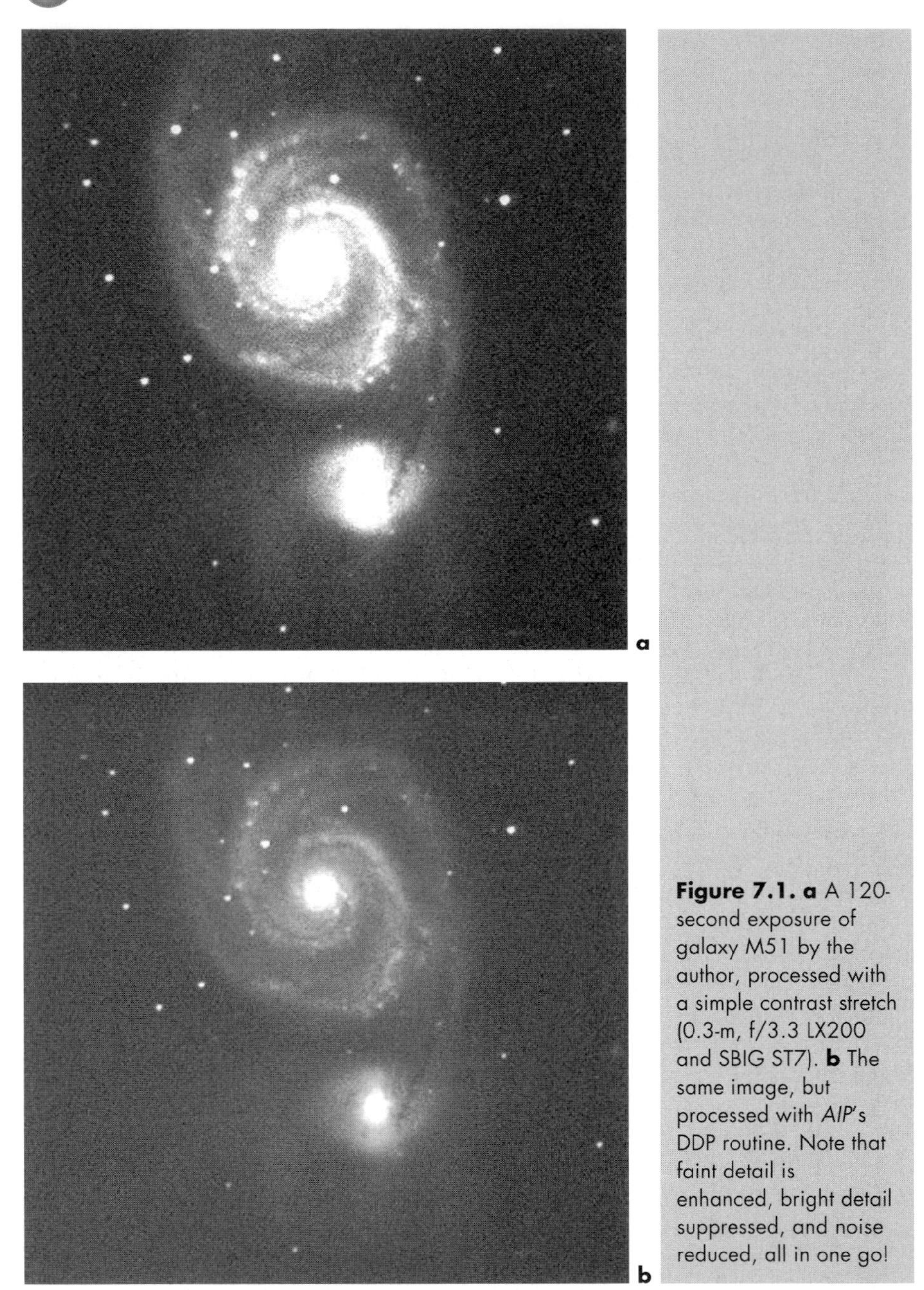

Figure 7.1. a A 120-second exposure of galaxy M51 by the author, processed with a simple contrast stretch (0.3-m, f/3.3 LX200 and SBIG ST7). **b** The same image, but processed with *AIP*'s DDP routine. Note that faint detail is enhanced, bright detail suppressed, and noise reduced, all in one go!

and 7.2) which, with a few key strokes, enhances faint spiral arms, reduces noise, prevents overexposure of the galaxy core, and reduces star images to mere pinpoints – superb! This worked fine with my old SBIG ST7 images, although I initially had a problem importing my Starlight Xpress 16-bit FITS files. All the bright stars had black holes in the middle when imported into *AIP*, but the problem was cured by saving the original as an 8-bit image. *AIP* also has a feature called a

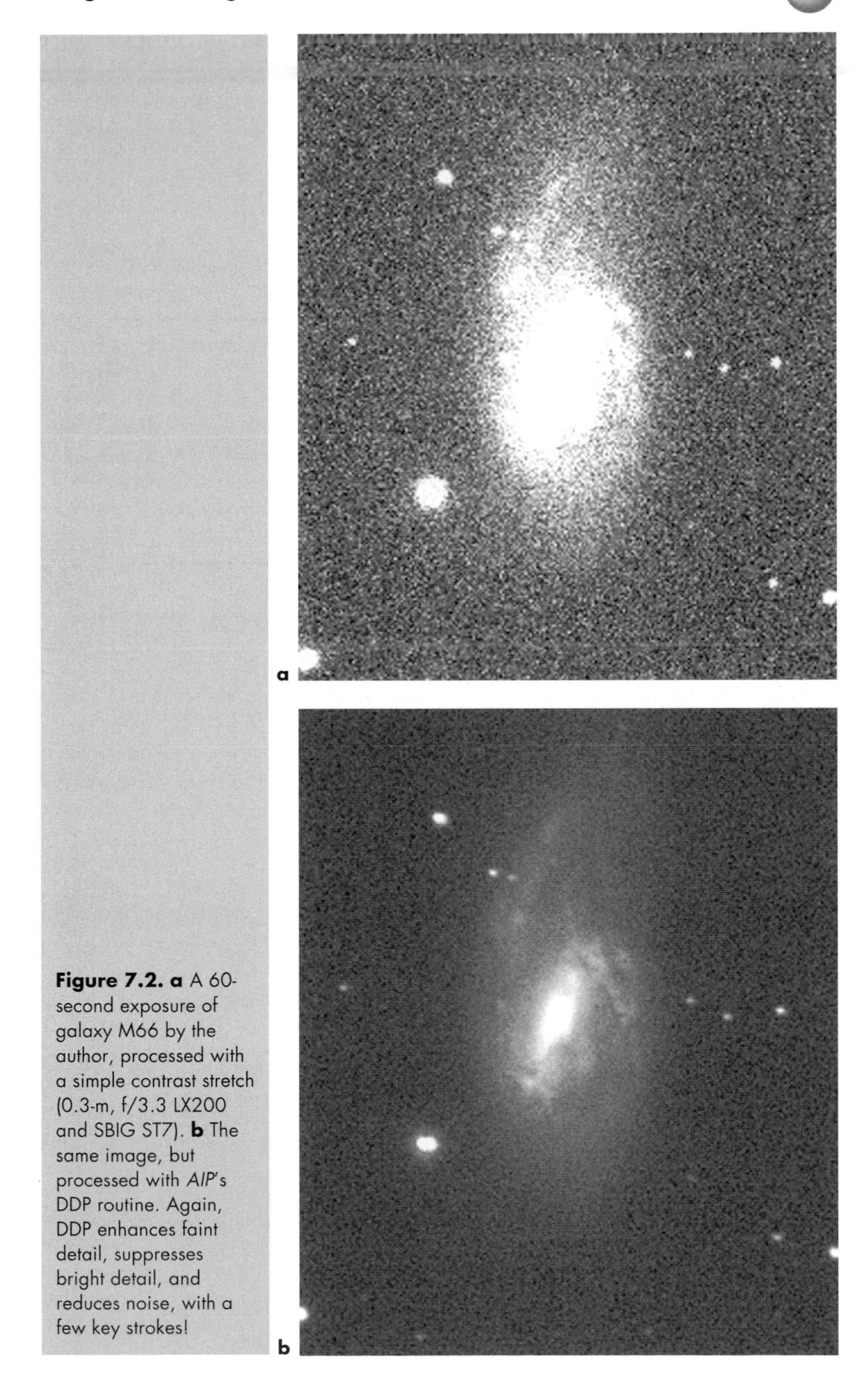

Figure 7.2. a A 60-second exposure of galaxy M66 by the author, processed with a simple contrast stretch (0.3-m, f/3.3 LX200 and SBIG ST7). **b** The same image, but processed with *AIP*'s DDP routine. Again, DDP enhances faint detail, suppresses bright detail, and reduces noise, with a few key strokes!

Universal Loader which means you can load any image and vary any parameter (e.g., row length, column length) while you do it.

Another example of the usefulness of *AIP4Win* is that it is one of the very few software package I know of that features a multi-image photometry tool. In other words, the software can automatically measure the brightness of a variable star in a whole stack of images and output a light curve! While I've seen custom home-made packages do this, I've never seen it done with ease in a commercial package.

As a final severe test on first acquiring *AIP*, I tested the software to see if it would import a friend's Starlight Xpress MX5 CCD nova patrol images and align and blink them. This was another test that all other packages had failed on; period! It could not have been simpler: *AIP4Win* loaded the files and then, after clicking on two different stars in both master and patrol images – wham! – the images were autoscaled, rotated, aligned, and were blinking; an awesome facility for any patroller.

As far as I can see, if you can change your file formats to FITS, or if your CCD camera's raw files are catered for by *AIP4Win* (very likely) you will not be disappointed by the *AIP* software from Willman–Bell. All in all, software and book, a very impressive combination.

Maxim DL and *DL/CCD* from Cyanogen

Maxim DL and *DL/CCD* (the latter controls the camera too) is a quality independent astronomical image-processing package. Like Software Bisque, Cyanogen Productions don't make CCD cameras; they are a Canadian astrosoftware company and have strong links with the CCD manufacturers. *Maxim DL/CCD* is also the most heavily advertised package, though, in this case, the glossy hype is no exaggeration. In terms of "value-for-money', *AIP* beats it hands down; plus, you don't get a 600-page hardback book explaining image processing with *Maxim DL*. What you do get is a very powerful, comprehensive, and intuitive image-processing package and a hefty user's manual. *Maxim DL* and *DL/CCD* are constantly evolving commercial packages. As more CCD cameras are released by the major manufacturers, *Maxim* add upgrades to their software and their Web page so that purchasers can upgrade for free, up to one year after purchasing the software. If you purchase the full *DL/CCD* option you will have a package that can totally control the CCD camera and numerous accessories as well. *Maxim DL* is ASCOM (Astronomy Common Object Model) compliant, which means that advanced users can use scripts to integrate *Maxim* with other software packages they are using. Among the numerous features included in *Maxim DL*, the following are particularly worthy of mention:

- Maximum entropy deconvolution;
- Digital development processing (DDP);
- Unsharp masking;

- FFT and kernel filters;
- RGB, CMY, and L (luminance) combining;
- Pixel math operations;
- Photometry and light-curve plotting;
- Image stacking, alignment, and blinking (similar to *AIP*'s tools);
- *Maxim*'s high bit depth ensures detail is rarely lost even after numerous processing steps.

Maxim's CCD camera control module currently has compatibility with all of the following CCD cameras/instruments:

- Apogee
- Astrovid
- Celestron
- FLI
- HiSIS
- Meade
- SBIG
- Starlight Xpress
- Most popular filter wheels, autoguiders, and SBIG's AO7 Adaptive Optics Units.

AstroArt 2.0

AstroArt is a much less expensive package than *Maxim DL* and, not surprisingly, has fewer features. However, it's definitely a strong option for the CCD user who feels that the software supplied with the CCD camera could be better. In terms of price, it's nearer to Berry's *AIP* book plus software, but Berry's impressive book may well swing the decision for most people. The software is compatible with most Starlight Xpress, SBIG, Apogee, HiSIS, LISAA, and Cookbook cameras but I found it especially adept at processing Starlight Xpress images, especially raw images from the one-shot color cameras. When it comes down to the "most wanted" requirements of a software package for amateurs, stacking images is always high on the wish list. For example, stacking images of a fast-moving comet, by using the comet's central coma as the reference point, is one desirable feature that every CCD user wants. *AstroArt* doesn't disappoint in this respect; it has an easily learnt image-stacking routine which is equal to *AIP* and *Maxim*'s systems but seems to run much more quickly (at least on my PC). Stacking planetary images to reduce noise is also accomplished easily. *AstroArt*'s other features include

- Easy to use Contrast/Brightness/Histogram functions;
- Unsharp mask, high- and low-pass, and median filters;
- Maximum entropy, Lucy–Richardson, and Larson–Sekanina (comet nucleus) filters;

CHAPTER EIGHT

CCD Planetary Imagers

When affordable CCD cameras first appeared on the scene in the late 1980s and early 1990s, amateur planetary observers were keen to appreciate their potential for shortening exposure times and "freezing" the atmospheric turbulence that is the bane of the astrophotographer. Indeed the SBIG ST4, the amateur astronomer's first commercial CCD camera, soon stopped being seen as an autoguider because it was more exciting as a planet imager! CCD cameras also have other advantages over film cameras as well as just raw sensitivity. They don't suffer from harmful shutter and reflex mirror vibration (although some do have moving vanes for facilitating dark frames); images can be viewed immediately; thousands of images per night can be taken; and, most important, images can be stacked and processed to bring out the maximum detail.

The Best Amateurs

Despite the fact that most planetary observers know about these advantages, few actually join the ranks of the world's best CCD imagers. Who are these people and why should they succeed where others fail? It may be instructive to list a few of the world's top planetary imagers and their equipment (Table 8.1).

As can be seen from the list, most of these observers use commercially available Schmidt–Cassegrains from Meade (30 and 25 cm) or Celestron (28 and 35 cm) and there is a bias towards SBIG CCD cameras, especially the ST5c. Throughout 2003, there was a distinct trend towards the Philips Toucam Pro webcam. In terms of location, two live in Florida, but the rest are scattered throughout the world. However, a more detailed inspection would reveal that

Table 8.1. A few of the world's best CCD imagers

Observer	Telescope	CCD	Location
Antonio Cidadao	25-cm and 28-cm SCTs	SBIG ST5c/FLI CM7-1E	Portugal
Ed Grafton	35-cm SCT	SBIG ST5c/ST6	Texas
Thierry Legault	30-cm SCT	HiSIS22/VestaPro Webcam	France
Don Parker	40-cm SCT and Newt.	SBIG ST9/ToUcam Pro	Florida
Damian Peach	30-cm and 28-cm SCTs	SBIG ST5c/ToUcam Pro	Tenerife/UK
Jesus R. Sanchez	28-cm SCT	ToUcam Pro webcam	Spain
Maurizio Sciullo	25-cm, f/8 Newt.	Starlight Xpress HX516	Florida
Eric Ng	25-cm and 30-cm f/6 Newts	ToUcam Pro webcam	Hong Kong
Tan Wei Leong	25-cm Takahashi Mewlon	ToUcam Pro webcam	Singapore

Cidadao, Parker, Ng, Wei Leong, and Peach have all observed from rooftop or balcony sites (see Figure 8.1); an interesting statistic! Another point worth noting is that all seven favor image scales in the region of 0.2–0.4 arc-seconds per pixel for their finest images (although image scales as low as 0.8 are sometimes used).

Using the formula already quoted in this book, with 10-micron pixels this translates to a focal length of:

$$206 \times 10 \text{ microns}/(0.2 \text{ to } 0.4) = 2060/(0.2 \text{ to } 0.4)$$
$$= 10{,}300 \text{ to } 5150 \text{ mm.}$$

So, with a standard 30-cm, f/10 SCT of 3000-mm focal length, a 2 or 3× Barlow lens is needed to increase the focal length. It is also fair to say that all these observers have locations which experience far better than average seeing conditions.

Download Time Is Crucial

When I first started writing this chapter in 2002, a few of the world's keenest amateur astronomers had started experimenting with webcams for planetary imaging. The main reason for this was that although many of them had CCD cameras like SBIG's ST5c, they lived in areas (like the UK) where poor seeing was the norm. Their planets looked like they were being viewed under a shaking bowl of water (and that's on a good night). Webcams have USB and even Firewire downloads. They are blisteringly fast at downloading images and some are almost as sensitive as astronomical CCD cameras, though all are much noisier. However, when a webcam can download 25 frames of Jupiter *per second*, you get far more opportunity to seize those perfect methods of good seeing. Unfortunately, individual webcam images are (in general) compressed (some data is lost) and they are noisy. What is needed is a software package which can easily be used by the operator to search through hundreds of frames of a webcam AVI video, select the best ones, and stack them up. This would reduce the noise considerably and result in an instant, quality, bitmap image, much smoother than the original. Well, amazingly, such a software tool does exist and it's freeware, courtesy of one Mr Cor Berrevoets!

Figure 8.1. a The legendary Don Parker and his superb, customized 16-inch (40-cm), f/6 Newtonian on its balcony location at Coral Gables, Florida. Don is one of the few observers whose excellence in planetary imaging spans both the photographic and CCD eras. Photo: courtesy Don Parker. **b** The 11-inch (28-cm) Celestron of Damian Peach sited on a terrace above Puerto de la Cruz, Tenerife, equipped with an SBIG ST5c CCD camera. Photo: courtesy Damian Peach.

His software, called *Registax*, works like a dream and can be downloaded from http://aberrator.astronomy.net/registax/

Many amateurs with less-than-perfect seeing are now using webcams alongside Mr Berrevoets' software and getting truly amazing results. In general they use a specific webcam, namely the Philips ToUcam Pro which uses a Sony ICX098AK interline CCD with 640 × 480 pixels. The chip's active area measures roughly 3.9 × 2.8 mm and has tiny 5.6-micron pixels, making it ideal for planetary imaging at f-ratios of 20 or so (i.e., a Schmidt–Cassegrain plus a 2× Barlow lens). The Sony chip is extremely sensitive and so ideal for planetary targets where exposure times need to be kept short.

The UK's Damian Peach, who normally uses an RGB filter wheel on an ST5c, has obtained superb results with the ToUcam Pro webcam , but has concluded that under good conditions, his ST5c is still a better performer. Its noise is so much lower and, with the infrared rejection filters of his RGB filters, the colors are much truer. Nevertheless, one only has to see the stunning Jupiter and Saturn images of Hong Kong's Eric Ng and Singapore's Tan Wei Leong (both of whom use 25-cm aperture telescopes and the ToUcam Pro webcam) to realize that webcams are a serious instrument for the planetary imager, especially if he or she suffers from poor atmospheric seeing and does not want to pay so much more for the popular SBIG ST5c.

Why is the SBIG ST5c such a favorite for planetary work? Well, before the days of 2 GHz PCs and USB downloads, downloading CCD images was a process which took a significant amount of time. The fewer pixels a camera had, the quicker the download time, hence the popularity of the original 192 × 164 pixel SBIG ST4. At 0.3 arc-seconds per pixel, the largest planet, Jupiter (spanning 50 arc-seconds maximum), still only needs about 160 pixels to cover the planet's width, so even the ST4 is good enough, provided your telescope drive is excellent! Download time is absolutely crucial when imaging a fast-rotating planet like Jupiter. The giant planet has a circumference of some 450,000 kilometers (280,000 miles) and has a rapid rotation period of 9 hours 50 minutes at the equator. This translates to an equatorial rotation speed of 46,000 kilometers per hour (29,000 miles per hour).

At Jupiter's closest distance to Earth of 590 million kilometers (370 million miles) the equatorial rotation on Jupiter's meridian translates to a motion of 16 arc-seconds per hour (arctan 46,000/590,000,000) or 1 arc-second in less than 4 minutes of time. The modern planetary imager will stack numerous images of the planet to reduce noise and will take images in red, green, and blue (or cyan, magenta, and yellow) to create a color image. In addition, the vast majority of the images taken will be unusable due to the poor seeing typically created by courtesy of the Earth's atmosphere. It doesn't take a genius to work out that with the planet's features moving at 1 arc-second in 4 minutes, a requirement for arc-second resolution and a need to take many images, the planetary imager has a real problem taking enough images before the planet has rotated too much. If more than a few minutes has elapsed between the first and last images taken, any attempt to align the images on obvious central features will result in a registration mismatch at the planet's edge, typically appearing as a colored fringe on tri-color images. If images are aligned by simply co-registering the planets disk, central features will be obviously misaligned if the planet's meridian has rotated by more than a pixel.

Figures 8.2 and 8.3a show images by Damian Peach and Ed Grafton of the giant planet; both observers used SBIG's ST5c. Figure 8.3b shows Ed Grafton's observatory.

Figure 8.2. Jupiter imaged by Damian Peach from a 7th storey balcony, with an SBIG ST5c CCD camera on 13 October 2000 at 0145 UT. A composite of dozens of luminance and RGB images taken over a 2-minute period using a 30-cm, f/10 Meade LX200 and Barlow lens, giving an f-ratio of approximately 21 and an original sampling resolution of 0.3 arc-seconds per pixel (10-micron pixels). Damian was based in Norfolk, UK, at this time. Photo: courtesy Damian Peach.

You can now see why SBIG's ST5c has long been a favorite with planetary imagers. With only 320 × 240 pixels and a resultant image download time of less than 3 seconds, good seeing provides an opportunity to obtain dozens of sharp images before planetary features have moved too much. Of course, with so many cameras switching to USB downloads in 2002, image capture times have plunged drastically. For example, SBIG's super-sensitive ST-9E 512 × 512 pixel parallel port camera used to have an image download time of 11 seconds. Its USB equivalent, the ST-9XE, has a download time (with a modern PC) of under a second (400,000 pixels per second) , a dozen times faster! It can now be used as a supernova patroller's camera and a planetary imager's camera – useful! But the ST5c has other advantages; it is inexpensive ($900 in 2002), has an optional, low-cost, tri-color filter wheel, and comes with superb software (*CCDSoft* and *The Sky* from Sofware Bisque and *CCDOPS/Planet Master* from SBIG). It also has small, 10-micron, pixels which are well-suited to the focal lengths achieved with Schmidt–Cassegrains coupled to a Barlow lens, at least, where planetary imaging is concerned. The ST-9XE is a much more expensive camera, but a much more quantum-efficient one too. In addition, anyone who has battled to keep a planet's disk on a tiny CCD chip will appreciate a few more pixels to accommodate the periodic error of the telescope drive. One extra feature of the ST5c that is well worth mentioning is its partial frame mode that downloads up to three frames per second if only a fraction of the chip is being used. In the UK, Damian Peach (Figure 8.4) has used this mode to take some amazing high-resolution shots of

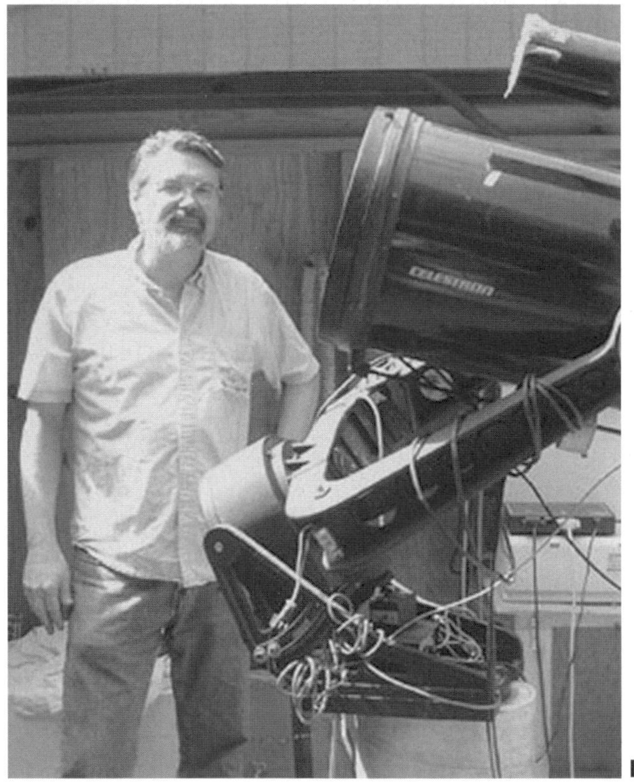

Figure 8.3. a Jupiter imaged by Ed Grafton from Houston, Texas, with an SBIG ST6 CCD camera on 2 October 2001 at 10:58 UT. The Great Red Spot and the shadow of the moon Io can be seen. This was a composite of dozens of luminance and RGB images taken using a 35-cm, f/11 Celestron 14 and eyepiece projection to f/66. Such a combination gives a sampling resolution of approximately 0.22 arc-seconds per pixel. Photo: courtesy Ed Grafton. **b** Ed Grafton and his Celestron 14, in his run-off roof observatory near Houston, Texas. This equipment has been used to take some of the highest-resolution planetary images ever obtained from the Earth's surface. Note the low angle of the polar axis at latitude 30 degrees north. Photo: courtesy Sheldon Faworski and Ed Grafton.

detail on Jupiter's Galilean moons (see Figures 8.5 and 8.6). When one considers that the opposition diameters of Io, Europa, Ganymede, and Callisto are 1.2, 1.0, 1.7, and 1.6 arc-seconds respectively, this is an amazing achievement!

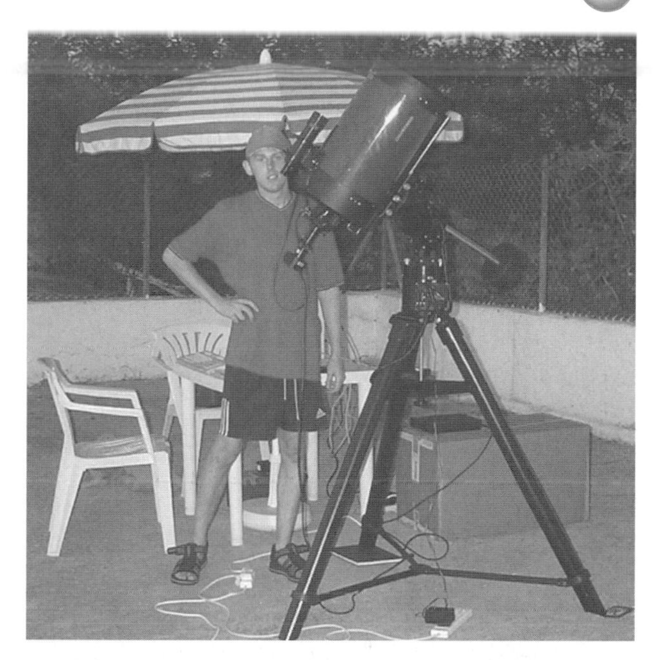

Figure 8.4. Damian Peach with his Celestron 11 Schmidt–Cassegrain at Puerto de la Cruz, Tenerife, in October 2002. Damian is, without doubt, the most prolific high-quality imager of the planets at the time of writing. Photo: courtesy Damian Peach.

Figure 8.5. Jupiter, its moon Ganymede, and the shadow of its moon Callisto. Imaged by Damian Peach with his 30-cm LX200 on 16 February 2002 at 19:40 UT. Jupiter's diameter was only 43 arc-seconds at the time. A composite of dozens of luminance and RGB images taken over a 2-minute period using a 30-cm, f/10 Meade LX200 and Barlow lens, projection to f/29.1 giving an original sampling resolution of 0.23 arc-seconds per pixel (10-micron pixels). Exposure times for each image were 0.15 seconds for the unfiltered luminance frames and 0.4 seconds for the filtered color frames with an SBIG ST5c camera. Damian was based in Rochester, Kent, UK, at this time. Photo: courtesy Damian Peach.

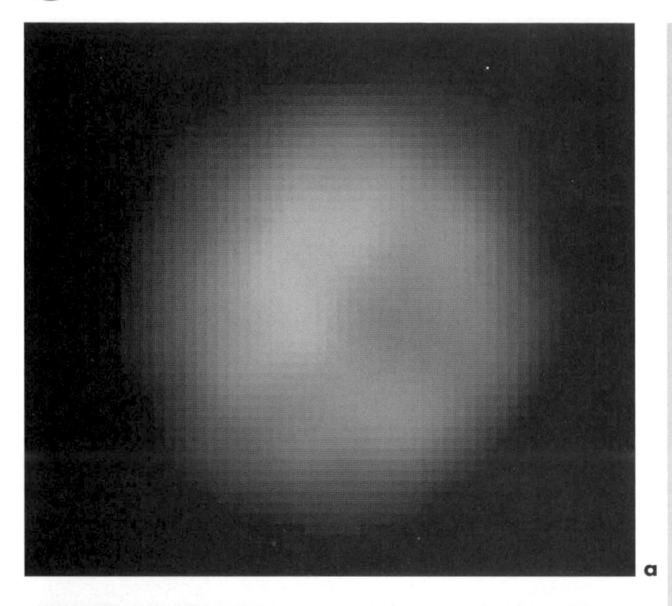

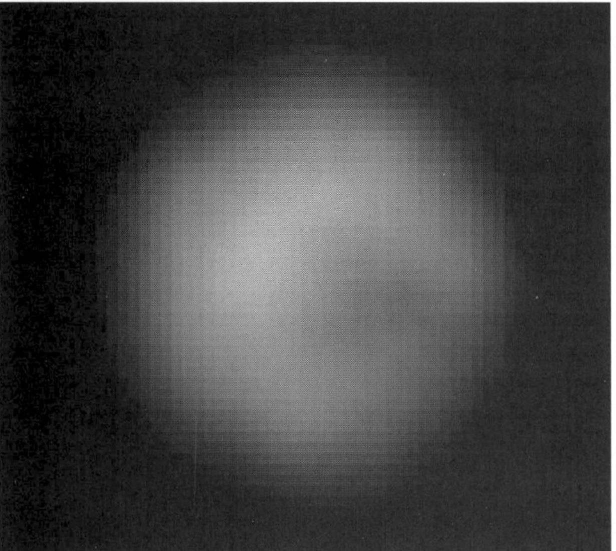

Figure 8.6. a Enlargement of Jupiter's moon Ganymede by Damian Peach. Details are the same as for Figure 8.5. Photo: courtesy Damian Peach. **b** Enlargement of Jupiter's moon Ganymede by Damian Peach. Details as for (a) except the image is 20 minutes later, at 20:00 UT. Photo: courtesy Damian Peach.

Different Planets: Different Rotation Speeds

While on the subject of Jupiter's rotation speed, we might as well look at the corresponding speeds for Mars and Saturn, and Jupiter's Galilean moons, lest the book reviewers say I've left something out! Mars has a circumference of 21,340 km, rotates in 24 hours 37 minutes, and can approach Earth within

59 million kilometers. Saturn has a circumference of 379,000 km, rotates in 10 h 14 minutes (equatorial), and has a minimum distance from Earth of 1200 million kilometers. The corresponding maximum (opposition) rotation speeds of the planet's meridians, as seen from Earth, are 3 arc-seconds/hour for Mars and 6.4 arc-seconds/hour for Saturn, so compared to Jupiter's 16 arc-seconds/hour, these planets allow far more time to collect those vital images. Jupiter's large Galilean satellites and their shadows have similar motions across the face of the planet as the planet's meridian features. Anyone who enjoys observing Jovian satellite and shadow transits will have noticed that when Io, Europa, or Ganymede are transiting the central part of the giant planet's disk, they appear almost to be part of the planet: atmosphere and satellite appear to move at almost the same rate.

The most frequent transiting body is Io as it orbits in 1.77 days at 422,000 km from the planet. The corresponding figures for Europa, Ganymede, and Callisto are 3.55/671,000, 7.16/1,070,000 and 16.7/1,880,000. At mean opposition distances from Earth these four orbital speeds correspond to motions of 20, 16, 12.7, and 9.5 arc-seconds per hour, so, in a 3-minute imaging span at opposition, Io will drift 1 arc-second and slowly overhaul the Jovian features beneath it. Of course, if any of the satellites are traveling over regions near to the planet's limb, their relative motion will appear faster due to foreshortening or the planet's smaller diameter at the poles. The fact that the satellites have a similar motion to Jupiter means that they can often be used for aligning images too, whether or not they are crossing the Jovian disk.

Matching Image Scale to Seeing Conditions

I have already mentioned that planetary imagers prefer image scales in the region of 0.2–0.4 arc-seconds per pixel, but, in practice, impressive planetary images can be taken with coarser resolutions, as low as 0.8 or even 1 arc-second per pixel. We have seen that the Dawes limit for the theoretical maximum resolution of an optical telescope is given by the formula:

Resolution in arc-seconds = 116/Aperture in millimeters.

So a 116-mm aperture will theoretically resolve 1 arc-second and a 232-mm aperture will resolve 0.5 arc-seconds. Due to a digital sampling law called the Nyquist sampling theorem, we need to sample at twice the frequency of the incoming data if information is not to be lost. In this context, we must sample at 0.2 arc-seconds per pixel if we hope to resolve to 0.4 arc-seconds, the theoretical resolution of a 290-mm aperture. However, even the most spectacular amateur images rarely resolve objects much smaller than one arc-second! Matching image scale to your local seeing is an important part of planetary imaging. Increasing the telescope focal length to give higher sampling will make the image dimmer and necessitate longer exposure times. Longer exposures give the Earth's atmosphere more chance to blur the picture. Keeping the exposure short and putting up with a very dim image is not a good idea; although brightness and contrast can be adjusted by image processing, a dim image will have a poorer signal-to-noise ratio than a bright one and a well-exposed image, filling at least half of the dynamic range of the camera, is strongly recommended.

Quite often, an image will show reasonable resolution but will look too blocky because a short focal length was needed to reduce exposure times. The Cassini division in Saturn's rings is a good example of a high-contrast feature which will look overly pixellated when long focal lengths can't be used. This is because, with a width of 0.7 arc-seconds at best, it only spans one or two pixels. A standard trick to make an image look more pleasing is resampling the image, that is, using a software package like Adobe *Photoshop* or *Paint Shop Pro* to increase the number of pixels in the image, while authentically retaining all the original data. No more information will be contained in the final image, it will just look better! Resampling is of critical importance for the modern planetary observer, especially when stacking images, as we will see later on.

So, what is the best strategy for matching your image scale to the local atmospheric seeing? As a rough guide I would advise the novice to start with a scale of about 0.8–1 arc-seconds per pixel and keep exposure times to about 0.1 seconds (for Jupiter) and then work to longer focal lengths from there. As you start to get better seeing you will naturally progress to longer and longer focal lengths until, if you get excellent results, you stop at around 0.2 arc-seconds per pixel (if you don't give up first!) Without a doubt the indifference of the Earth's atmosphere to the needs of the planetary imager is the main reason why most amateurs never manage to take good images of the planets. It is all too easy, after months or years of trying, to throw the towel in and accept that you will never get quality planetary images. If you use a CCD camera at a short exposure time (e.g., one hundredth of a second) and take images of a bright star, it is horrific to see the contortions that the star's disk goes through in successive images; all due to the Earth's turbulent atmosphere!

While seeing is, perhaps, the number-one problem, in reality the planetary imager must systematically remove *all* the obstacles in his (or her) path which are blocking the track to imaging perfection. I can't stress this aspect enough! All of the following issues must be tackled:

- Atmospheric seeing, local turbulence, and weather forecasting;
- Keeping the optics clean;
- Collimating the optics;
- Taking flat-fields;
- Focusing the telescope/CCD system;
- Stacking images precisely and reducing noise;
- Perfecting image-processing techniques.

Let's now examine all these points in much greater detail!

Seeing

Is there anything that can be done about the atmospheric seeing, or the local seeing? The world's leading planetary imagers have found that hilltop locations or balcony observatories often experience superior seeing. This is because heat, rising from the cooling night-time ground, causes severe turbulence the closer you are to ground level. The altitude of a planet is also a vital consideration; poor seeing is far more likely if a planet is low down. As a general rule (but there are exceptions) high resolution will rarely be achieved when a planet's altitude is less than 40 degrees. If you live in the UK (average latitude 52 degrees north) this means the planet must transit at a declination above +2 degrees and preferably much higher! Still night-time air, almost always associated with a large high-pressure air mass, will also auger well for good conditions. Indeed, when the weather forecast predicts misty or foggy conditions, this will generally mean good seeing. I have had some of my best planetary views when a mist was forming, with a thick dew on the telescope tube. After sunset, temperatures on a clear night will fall rapidly at first and then start to fall much more slowly after a few hours. The dawn skies, even in quite significant twilight, can often bring the most stable atmospheric conditions. Thus, good seeing is often experienced more often before a planet reaches opposition, rather than afterwards. Many serious planetary observers have also sought ways to improve their local seeing conditions by using vibrationless fans to create a laminar air flow (in Newtonian tubes) and to cool the mirror down to the same temperature as the night air. Turbulence caused by heat rising from a thick glass mirror is a major source of poor planetary definition. On this subject, a dome is *not* a good idea for the planetary observer; heat builds up inside a dome during the day and streams out of the dome slit at night. The nine observers listed in Table 8.1 do not use domes, they use balconies, roof-tops, or open-air observing. In practice, most observers will rarely get perfect seeing, even if all of the above measures are implemented. However, even without perfect seeing, the new amateur astronomer has one very powerful weapon. Even in mediocre seeing, he or she can take thousands of short-exposure images per night, maximizing the chance of a few being sharp. If this is combined with observing on as many nights as possible, throughout the apparition, a few exceptional images *will* result, providing the other issues below are addressed. Another advantage CCDs bring the planetary observer, is that he or she, can be positioned well away from the telescope; it may seem a trivial point, but heat from the observer's body is a significant source of local turbulence and poor seeing.

Cleaning Optics

Cleaning telescope optics is not something to be entered into without considerable planning and the utmost care. This is one area of telescope maintenance where prevention is most certainly better than cure. While it is true that a mirror can look revoltingly dirty from corrosion and accumulated condensation stains, but still give pleasing views, if we are trying to take high-resolution planetary images we want the optics to be as clean as possible. Remember, good optics are accurately figured to a precision much better than a tenth of a thousandth of a millimeter, so any transparent deposit will make the telescope perform as if it was less accurately figured. One of the major hassles of amateur astronomy is the fact that precision optical instruments are taken outside into an environment where there is plenty of dew, dirt, dust, and insects about. There could be no dirtier environment for any telescope to operate in than someone's back garden! This is where the planetary balcony, or patio, observers with semiportable Schmidt–Cassegrains have a distinct advantage: they can wheel, or drag, their precision instruments back indoors when an observing session is over. While a few amateur astronomers' living rooms are as dirty as their gardens, most are not!

The culprit in nearly all cases of dirty, stained optics is condensation. Condensation is often kept at bay with dew heaters, but sources of heat are not wanted for planetary observing: everything should be kept in thermal equilibrium. So what do the experts do? Many amateurs keep low-wattage greenhouse heaters or light bulbs in their Newtonian tubes to keep the damp at bay when the telescope is not in use. Others use tightly fitting dew caps that fit over the main mirror or corrector plate and allow a minimum of air to be trapped between cap and optics. In night-time use, a long dew shield is recommended for Schmidt–Cassegrains or the corrector plate will soon dew up. But heated dew shields are only recommended for deep-sky observing. Silica gel is a substance that readily absorbs moisture and there are even variants where the granules change color (usually from blue to pink) when damp. But silica gel needs to be used sensibly. Leaving bags of it in an open telescope tube will result in the bags being saturated with condensation after one night! The trick is to cap unexposed optics such that a very small amount of air is trapped above the mirror/corrector plate and then insert a sachet of silica gel in the gap. The gel should be dried out in an oven every week if at all possible.

If optics are allowed to get dirty then, unless they are due for realuminizing, you may want to clean them yourself. Personally, this is something I tend to avoid (and would advise you to, as well), but if you really have your heart set on cleaning your main mirror yourself here is what to do:

1. Remove the mirror from the telescope; don't touch the mirror surface with your fingers (and don't drop it!).
2. Place the mirror, reflective surface up, in the kitchen sink and let a gentle stream of warm water flow onto the center of the mirror, to remove loose clumps of dirt. Do *not* use hot water as the thermal shock may crack the mirror in half! If the mirror is thin, or plate glass, even warm water could crack it if the mirror is cold.
3. Place some drops of detergent all over the wet surface of the mirror and leave for ten minutes to loosen stubborn dirt.
4. Then rinse the mirror with tap water for several minutes.
5. Carry out a final rinse with distilled or demineralized and well-filtered water.
6. Position the mirror on its edge and don't let it roll off the sink on to the floor! (Don't laugh – it *has* happened). Let the water drain off, but when persistent droplets start to stick to the surface, blow these off with an air-blower aerosol (electronics stores sell these) before they dry and leave a mark.
7. Re-install the mirror.

Collimating the Optics

Without a doubt, most amateurs who own Newtonian reflectors, Cassegrain reflectors, or Schmidt–Cassegrains fail to collimate them properly. At a stroke, this renders them incapable of high-resolution planetary imaging. Refractors and Maksutovs have, pretty well, factory-fixed collimation, and this is often why they are considered to be such good planetary instruments. Collimating a telescope is not difficult, but it does require a systematic and patient approach. Newtonians and classical Cassegrains are collimated by adjusting both the main and the secondary mirrors. Commercial Schmidt–Cassegrains can only have their secondary mirrors adjusted for collimation.

Daytime Collimation

Figure 8.7 shows the three phases in the basic daytime collimation of a Newtonian telescope. They show the view, through the drawtube, when the observer's eye is centered where the eyepiece normally sits. A plastic 35-mm film canister with a hole in the middle makes an excellent sighting tube for positioning the eye. The next step is night-time star collimation. With a Schmidt–Cassegrain, there is only one daytime step, i.e., adjust the three screws on the secondary mirror until the reflection of the primary in the secondary is concentric. For both Newtonian and Schmidt–Cassegrain collimation, the job is a hundred times easier if an assistant is employed, so the telescope owner can look through the collimating eyepiece while the assistant adjusts the mirrors. The task of Newtonian collimation is also much easier if a laser collimating eyepiece is purchased, or one of the many collimating tools available from companies like Kendrick. Up to a year or so ago, collimation tools were only available for Newtonians, but SCT collimation tools are also now available. How to use these tools is beyond the scope of this chapter, as upon purchasing them you will be provided with precise instructions. However, even with these tools, all advanced amateurs will want to fine-tune their telescopes by collimating them on a star.

Star Collimation

The first stage in star collimation is probably unnecessary if careful daytime calibration has already been carried out. It simply involves observing a first-magnitude star at about 250×, and well out of focus, and checking that the black hole (the shadow of the secondary mirror) is in the middle of the star's out-of-focus disk. Figure 8.8a shows the real image of an out-of-focus star, imaged with a well-collimated Schmidt–Cassegrain, by Damian Peach. Note that the real image looks a bit different from the perfect simulation in Figure 8.8b! Figure 8.8b, like most of the collimation images here, was produced using a simulation package called *Aberrator*. If the black hole is not in the middle, the mirror collimation screws need to be adjusted until the shadow is centered.

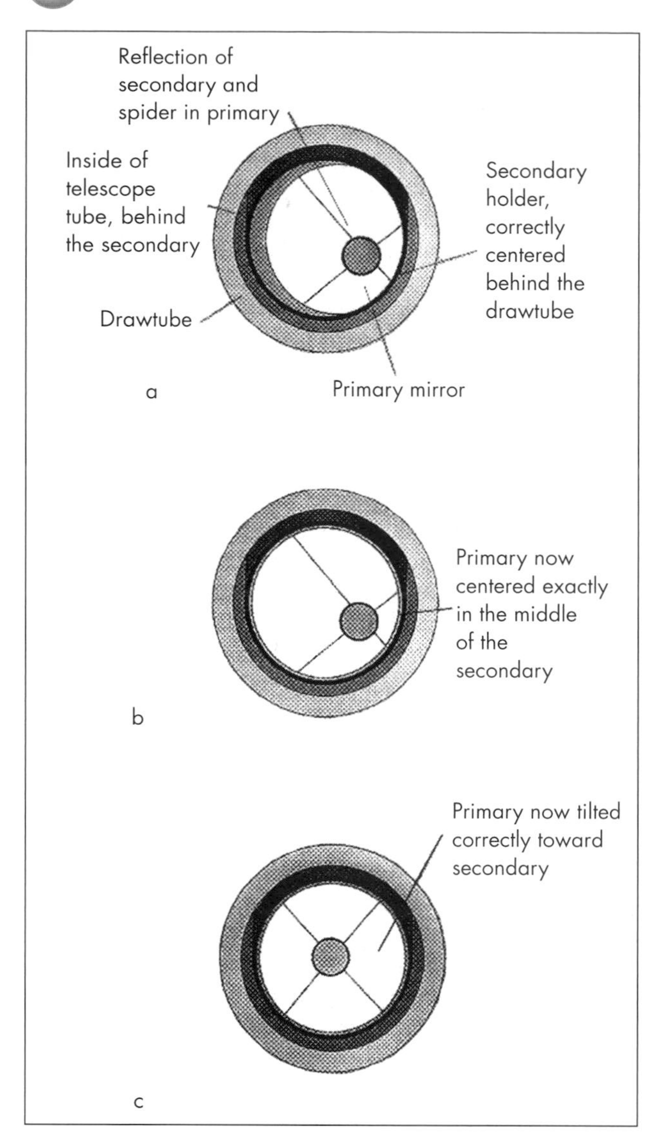

Figure 8.7. Three steps in initially collimating a Newtonian reflector, prior to star collimation – this is far less critical in a slow Newtonian.

The second stage in star collimation is far more demanding and can only be carried out when seeing conditions are reasonable. Typically, a third-magnitude star is chosen (with a 300-mm aperture) and it must be well above the horizon so turbulence is minimized. A very high magnification is then used (600× or more) and the star is moved from well inside to well outside of focus while the diffraction rings are examined. There should be a bright dot in the middle and then a series of concentric dark and light rings out from the center. As the scope is moved through focus this pattern should open and close smoothly and symmetrically (Figure 8.9); if it doesn't, the mirror adjusting screws need tweaking. Of course, every time the screws are tweaked, the star will move and will need recentering in the field. Once the intra- and extra-focal

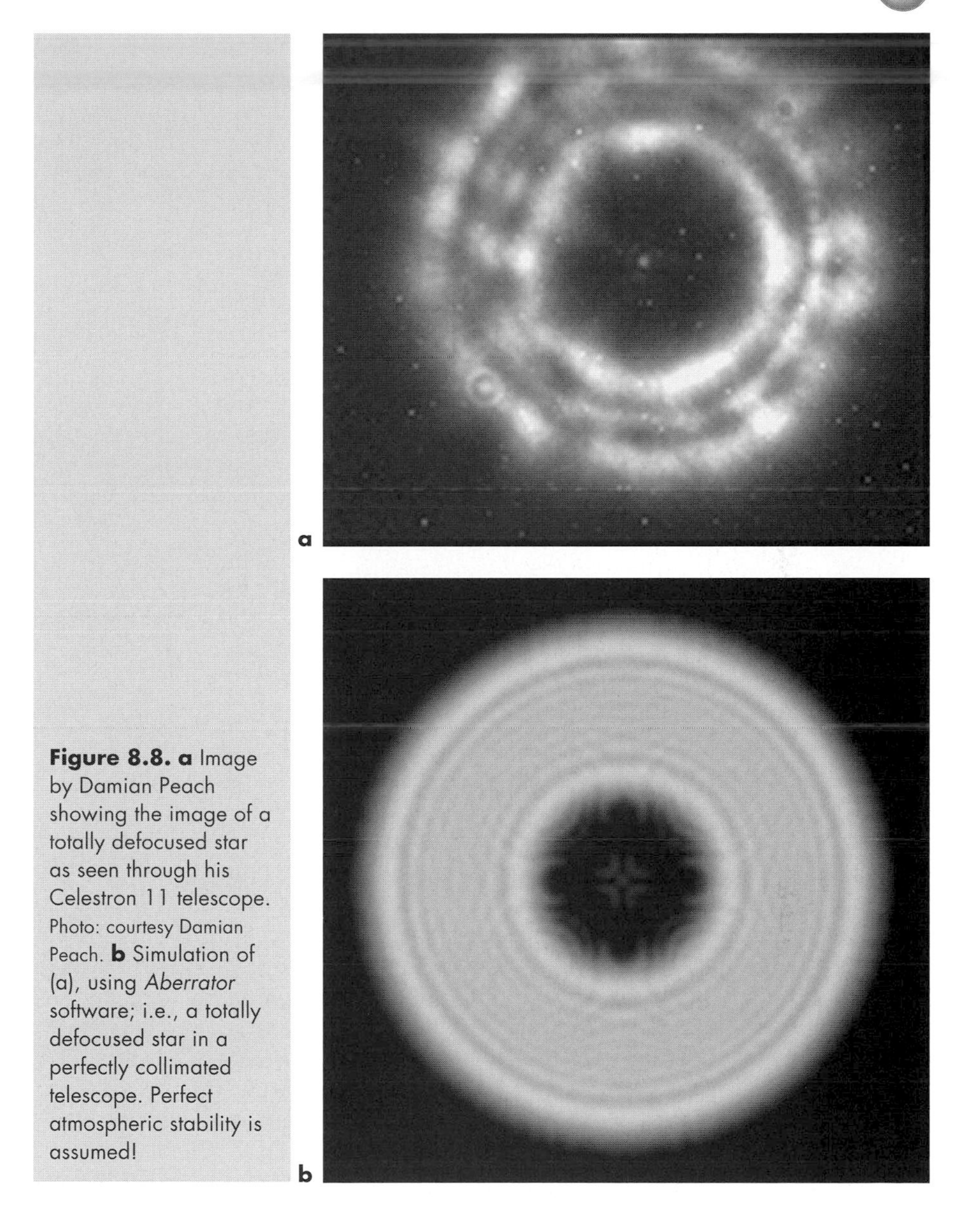

Figure 8.8. a Image by Damian Peach showing the image of a totally defocused star as seen through his Celestron 11 telescope. Photo: courtesy Damian Peach. **b** Simulation of (a), using *Aberrator* software; i.e., a totally defocused star in a perfectly collimated telescope. Perfect atmospheric stability is assumed!

patterns resemble a perfect textbook diffraction ring, the third step can be carried out.

The third and final step to perfect collimation can only be executed when seeing conditions are near-perfect – a rare event for most people. The set up is as for step 2, except that the star is perfectly focused. We are now looking for the perfect Airy disk, a so-called "false" central disk, surrounded by diffraction rings of diminishing brightness (Figure 8.10a). If the first ring is not uniform, or is incomplete (as in Figure 8.10b), the collimation screws need tweaking by a tiny amount to achieve a complete and uniform first ring. This last test is so sensitive

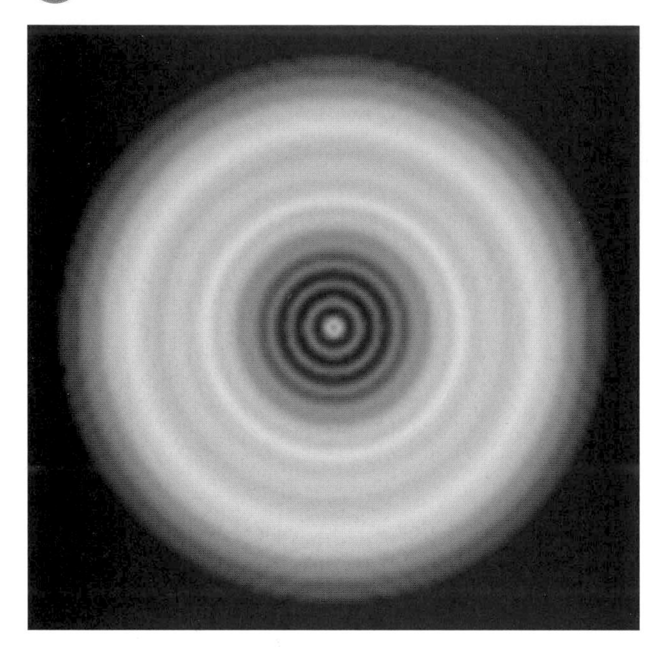

Figure 8.9. A simulated image of a slightly defocused star in a perfectly collimated telescope.

that even moving the scope around the sky will alter the situation! Hard-core planetary imagers will often tweak their scope's collimation with this final step, before they commence imaging and even adjust the collimation throughout the night! Are you still sure you want to image the planets?

Well, we've now covered collimation, but it should be emphasized that for the star tests, the diffraction patterns seen will not resemble anything like the textbook patterns unless the night is perfect. If you want to see textbook diffraction patterns on a less-than perfect night, try looking through a small-aperture quality Maksutov at a very bright star. This will enable you to familiarize yourself with what should be seen. Users of giant Newtonians, e.g., 40 cm and larger, will have to accept that they may never see a perfect diffraction pattern or Airy disk unless they stop the instrument down! Not only are the atmospheric "cells" of stable air rarely more than 30 cm across, large telescopes take a very long time to cool down.

Of course if, even on a good night when stars aren't shimmering wildly, the Airy disks look distorted, and you've been through the telescope collimation process carefully, you may have inferior optics. To be honest, this is very rare with optics from either Celestron or Meade or any of the major manufacturers. Competition is fierce in the modern telescope market and companies just can't risk turning out poor optics. However, if you are using an old telescope or one of dubious second-hand origin the optics might be suspect. However, in my experience, most amateurs who complain of poor optics simply haven't collimated them to the precision required to get diffraction-limited views of the planets. Also, a mirror cell that "pinches" the optics is often a source of non-perfect star images.

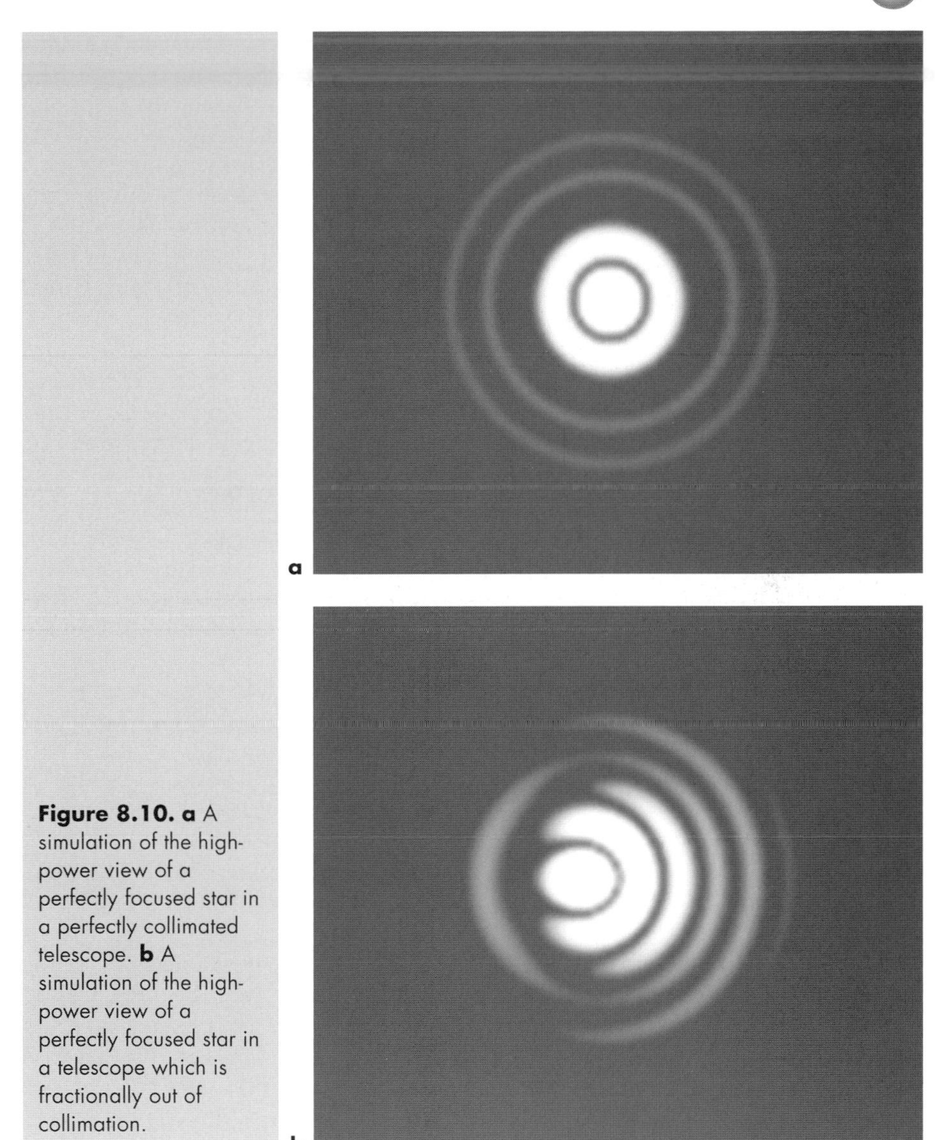

Figure 8.10. a A simulation of the high-power view of a perfectly focused star in a perfectly collimated telescope. **b** A simulation of the high-power view of a perfectly focused star in a telescope which is fractionally out of collimation.

Taking Flat-Fields

Planetary imaging is carried out at long focal lengths and long f-ratios. This means that the cone of light passing through to the CCD surface is very narrow. As a consequence, any specks of dust (or condensation stains) on the glass window, above the CCD chip, will obstruct a significant amount of the light heading for each pixel. When one takes into account the powerful image-processing routines (primarily unsharp mask ones) used by planetary imagers, the resulting "dust doughnuts" can ruin an otherwise pleasing image. One solution to this problem is to take quality flat-field images. A flat-field is simply the image of an evenly illuminated background, taken with the same instrument orientation and f-ratio as the main astronomical image. Such an image will contain a record of all the imperfections in the optical system (dust-specks, vignetting, etc.) and, when you divide the main image with the flat-field image, all the dust specks and uneven illumination will disappear. Well, that's the theory, and when it works well, the improvement is spectacular! A frightening "flat-field from hell,' supplied by Nick James, is shown in Figure 8.11. In this case the dust doughnuts are actually caused by frost starting to form on the CCD camera window. However, a bad flat-field can be worse than no flat-field at all. Most flat-fields are achieved by imaging the twilight sky such that it fills half the dynamic range of the CCD. This is best carried out as soon as the sky is dark enough for exposures of a second or so not to saturate the CCD chip. As the sky gets darker it will be harder for twilight to half-saturate the CCD and the longer exposures will start picking up stars, even if the drive is switched off! The problem is that we are talking about specks of dust here and specks of dust, by their very nature, move around from night to night and even minute to minute! So, although I would recommend flat-fields, it is yet another hassle to incorporate into the hectic planetary imager's schedule (along with nightly collimation and refocusing), but we are striving for perfection after all!

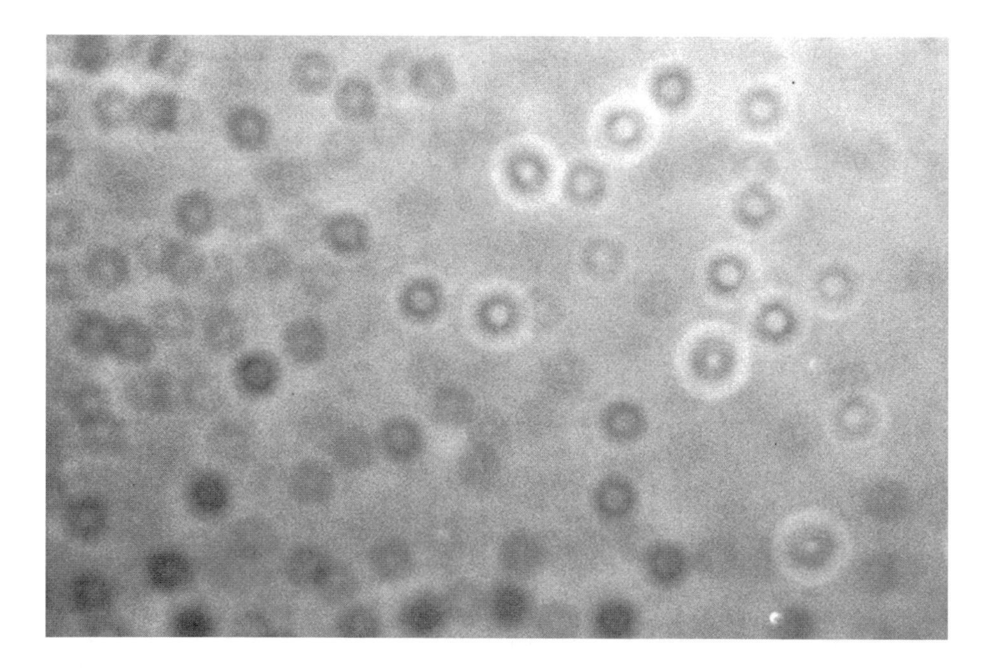

Figure 8.11. A flat-field from hell, caused by frost forming on condensation centers on the CCD faceplate. Image: courtesy of Nick James.

Alternative solutions to flat-fielding are:

1. Keeping the CCD faceplate scrupulously clean and avoiding imaging on the dusty bits; or
2. Combining so many quality images (as the planet image drifts around the CCD chip) that the dust speck problem averages out. Vignetting is very unlikely with narrow-field planet work, but dust and condensation specks are a bigger problem. If you stack dozens of images up, and if the planet drifts around on the chip, you may not need a flat-field. (A less-than-perfect telescope drive can be a bonus here!)

Focusing

It must be obvious that there is a major problem to be solved in simply focusing the telescope for planetary work; how do you focus on something that's constantly blurring in numerous different ways!

The technique that some planetary imagers use is to focus on a very bright star near to the planet first, using a very short-exposure time to minimize seeing effects. This focusing period may take half an hour before the imager is happy (by which time it will be cloudy of course!). Focusing on a bright star is highly recommended, but you may need to do it more than once per session, because as temperatures change and telescopes move, so the focus point changes. One aspect that is rarely appreciated is just how difficult it is to focus a telescope when the f-ratio is slower than about f/20. Schmidt–Cassegrains are a nightmare in this context, because focusing is usually achieved by moving the f/2 primary mirror nearer or farther from the secondary.

Let's have a look at some frightening statistics here. If we are using a 250-mm aperture reflector with an absurdly long f-ratio of 20, its diffraction limit of 0.46 arc-seconds will correspond to 0.011 mm at the focal plane. If the individual light cones from two points 0.011 mm apart are not to overlap we mustn't move in or out of focus by more than 20×0.011, or ± 0.22 mm. It turns out that this is independent of aperture because the diffraction resolution size at the focal plane (0.011 mm) will grow with reduced resolution but shrink with reduced focal length. In fact, the focus tolerance gets smaller with the square of the f-ratio measured at the moving focuser element in the system. This is a bit of a mouthful, but, basically, it means that focusing by moving the f/2 mirror in a Schmidt–Cassegrain will be 25 times more critical than using a focuser attached to the f/10 focus at the drawtube! The figure of ± 0.22 mm at f/20 becomes ± 0.055 mm at f/10, ± 0.014 mm at f/5, and ± 0.002 mm at f/2!

Imagine trying to adjust a mirror within 2 microns of the right position; well this is what is required with an f/2 Schmidt–Cassegrain if focusing by moving the mirror (see Figure 8.12)! It's easy to see from all this that the best place to put a focuser is at the end of the optical light path, i.e., just before the CCD camera. Not surprisingly, this is precisely what the world's leading planetary imagers do. Companies like Jim's Mobile Inc. (JMI) manufacture special focusers for use with Schmidt–Cassegrains, specifically because of the focusing problems encountered

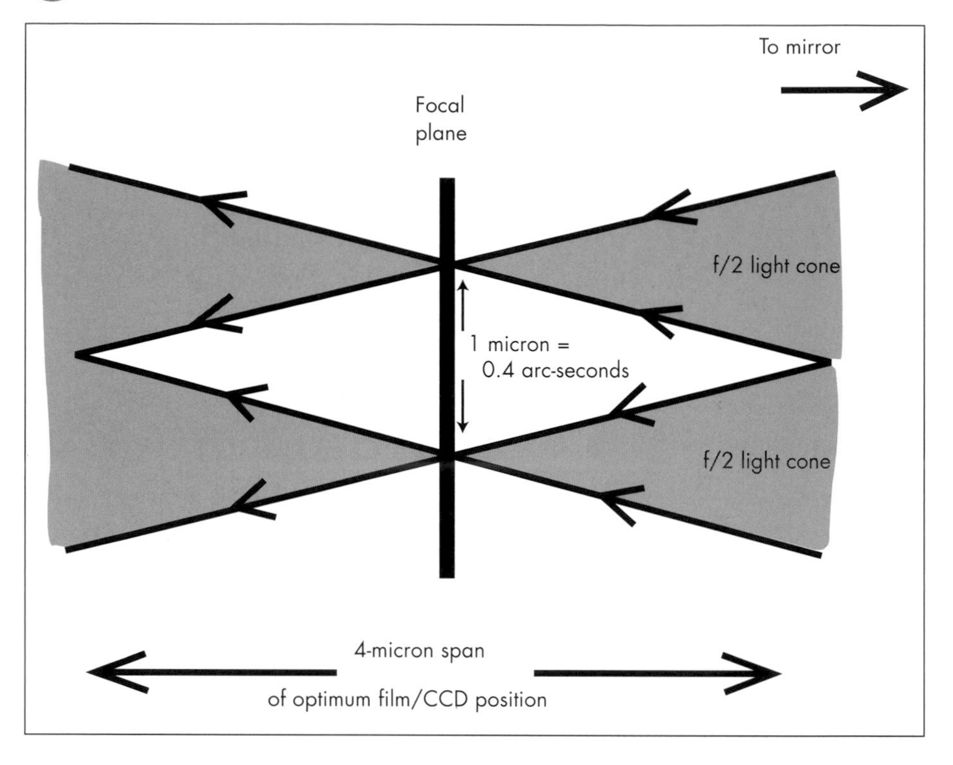

Figure 8.12. The focusing tolerance on an f/2 telescope is only ±2 microns if the full resolution is to be captured. In practice, planetary telescopes use much higher f-ratios, but, if the f/2 mirror in an SCT is moved to achieve focus, the focusing tolerance is just as critical.

at f/2. Their NGF-S (New Generation Focuser for Schmidt–Cassegrains) can be supplied with a motor, digital read-out of focus position (Figure 8.13), and even temperature compensation.

Other amateurs employ a variety of customized focusers and optical aids to focus their telescopes and keep the planets in view. The busy end of Don Parker's 16-inch (40-cm) Newtonian is shown in Figure 8.14. This may look somewhat cluttered, but just look at the results he achieves (Figure 8.15). In passing, I'd like to mention the legendary UK astrophotographer and optician Horace Dall (1901–1986) who photographed the planets at f-ratios of up to f/200 with his 39-cm Dall–Kirkham Cassegrain. At f/200 the focuser tolerance is a massive ±22 mm!

So we've now learnt a few tricks about focusing, but what if you don't want to keep constantly refocusing on Jupiter during an observing session? One trick I have used when imaging Jupiter is to find one of Jupiter's moons (Io is always within 2 arc-minutes of the Jovian limb) and to aim to get it as small as possible; the moons aren't point sources of course, but they are a good indicator. Before image processing, Jupiter's features are vague and of low contrast and judging focus is tricky. If a shadow of one of the Galilean moons is on the disk, this can be

Figure 8.13. The superb JMI NGF-S Motofocus unit, which allows smooth, accurate focusing, with a digital position read-out. Photo by the author.

used as a focus indicator too. At least with Saturn, the rings and the Cassini division are high-contrast features which can be used to advantage.

Stacking Images to Reduce Noise

If there is one powerful weapon that image processing has given the planetary imager it is the ability to stack multiple images to reduce noise. In all forms of imaging, a high signal-to-noise ratio will mean a higher-quality image. When very short exposures are used to freeze planetary detail the signal from the planet will be low and when the image is subsequently brightened and contrast stretched, noise will become apparent. If your single planetary images are faint and noisy because short exposures are the only way to freeze the seeing, try a shorter image scale (up to, say, 1 arc-second per pixel) and an exposure which will half-saturate the CCD. Stacking images allows an additional way of reducing noise, without relying solely on short image scales and long exposures. Noise can arise from a variety of sources and is often random; this is especially so when a planet's disk drifts around different pixels and different areas of the glass CCD window. Noise is also grossly exaggerated when powerful image-processing routines are used. However, noise reduces with the square root of the number of images stacked; thus, all other things being equal, a stack of four planetary images will have half the noise of a single image and a stack of 16 planetary images will have a quarter the

Figure 8.14. Don Parker's SBIG ST9E USB CCD camera, filterwheel, viewfinder, and customized Newtonian focuser on his 16-inch (40-cm) f/6 reflector. The ST9 is used at f/46 for an image scale of 0.22 arc-seconds per pixel. Photo: courtesy Don Parker.

noise of a single image. Some excellent examples of stacked planetary images are seen in Figures 8.16, 8.17, and 8.18. How many images do the masters of planetary imaging stack to get one good image and what exposure times do they use? Let us take the example of Damian Peach, who, unlike my other examples, has had to contend with the notoriously bad seeing of the UK, before he moved to Tenerife!

Damian Peach's Technique

Damian has used two instruments to obtain his remarkable results: a 30-cm Meade LX200 plus Barlow lens giving f/29.1 and a 28-cm Celestron plus Barlow lens giving f/31.4. Both systems deliver about 0.23 arc-seconds per pixel with the 10-micron pixels of an SBIG ST5c. For Jupiter, Damian typically takes LRGB

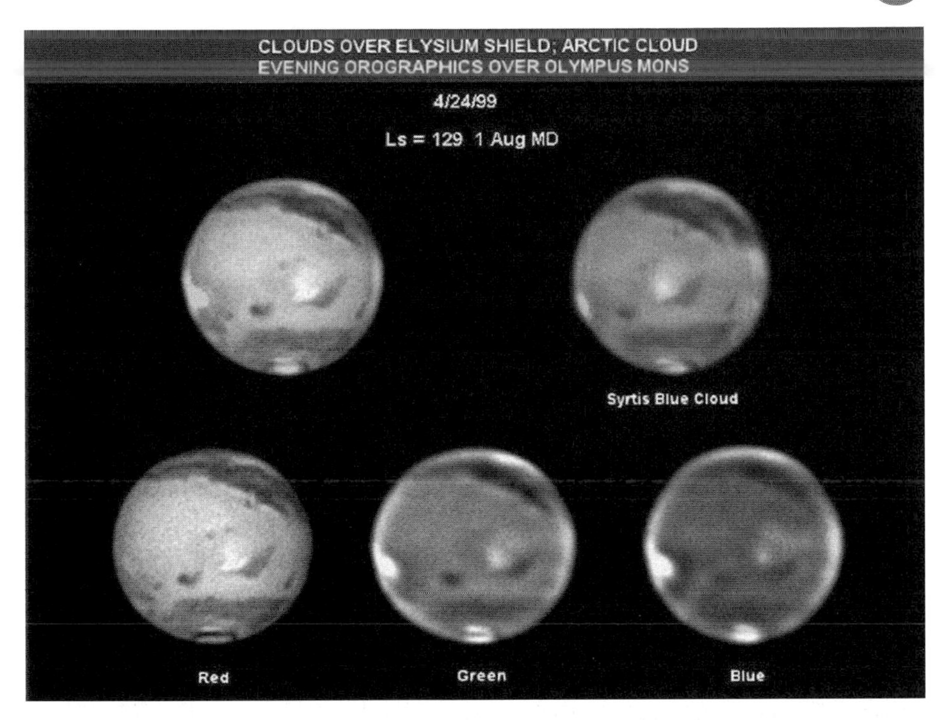

Figure 8.15. Mars imaged by Don Parker on 24 April 1999 with his 16-inch, f/6 Newtonian at f/48 and a Lynxx CCD camera. Filters used were: Blue Schott BG12+IR rejection; Green Schott VG9+IR rejection; Red RG610. Images were processed in Richard Berry's *BatchPix, ColorPix,* and *Photostyler.* The images show a "cold front" coming off the North Polar Cap, a brilliant orographic cloud over Olympus Mons on the evening (left) limb, and clouds over the Elysium volcanic shield near the center of the disk. Also note the famous Blue Syrtis Cloud on the morning limb in the top-right image, taken slightly later. Photo: courtesy Don Parker.

images where the luminance components are short unfiltered ones of 0.2-seconds exposure and the filtered red, green, and blue (with IR rejection) exposures are about 0.4 seconds each. On a typical night, Damian saves thousands of images at a rate slightly faster than one every 3 seconds and, during the night's best 3-minute period of good seeing, he can save up to 70 Jupiter luminance images for stacking. A typical sequence in a 3-minute period of excellent seeing might therefore consist of two sets of 35 luminance images and maybe six to eight images in each of the three colors, taken slightly later (the eye is less tolerant of color resolution). In each 90-second luminance span the features at the center of the Jovian disk will have moved a maximum of 0.4 arc-seconds (with the planet at opposition). Damian then stacks the resampled (much enlarged) two 90-second sets of images up and combines them, allowing for the planet's tiny rotation between the two sets of images. Finally, the RGB color component is added. The result is breathtaking!

With Saturn's slower rotation, Damian has found he can collect up to 80 luminance frames in a 220-second time-period, without worrying about planetary rotation.

Figure 8.16. Saturn on 18 October 2001 at 08:52 UT, imaged by Ed Grafton with a Celestron 14 working at f/68 and an SBIG ST6 CCD camera. A composite of dozens of luminance and RGB images taken over several minutes. Original sampling resolution, approximately 0.21 arc-seconds per pixel. Photo: courtesy Ed Grafton.

Figure 8.17. Mars on 11 June 2001 at 07:03 UT, imaged by Ed Grafton with a Celestron 14 working at approximately f/60 and an SBIG ST6 CCD camera. Photo: courtesy Ed Grafton.

Figure 8.18. Saturn on 17 October 2002 imaged by Damian Peach from Puerto de la Cruz, Tenerife, Canary Islands. 11-inch (28-cm) Celestron SCT at f/31.4. SBIG ST-5c camera with 10-micron pixels. SBIG RGB filter set. An average of an incredible 152 images over an 8-minute period were taken, comprising 119 luminance (0.35-sec) exposures, 10 red-filtered exposures (0.6-sec), 12 green-filtered exposures (0.6-sec), and 11 blue-filtered exposures (0.7-sec.) Unsharp mask processed. The equatorial diameter was 19.2 arc-seconds at the time and the planet 54 degrees above the horizon. Images were stacked in two sets to allow for the planet's rotation and minimize rotational "smear" to below 0.3 arc-seconds. Original sampling resolution: 0.23 arc-seconds per pixel. Photo: courtesy Damian Peach.

During 2003 Damian moved back from Tenerife to the poorer skies of the UK and switched to largely using the Philips ToUcam Pro webcam instead of his SBIG ST5c. His most recent results with Saturn have been generated by stacking up to 3,000 webcam frames, collected over a 300-second period. For Jupiter, he typically stacks up to 1,000 webcam frames collected in a 120-second imaging period.

It should, perhaps, be stressed again that the human eye and brain is far more perceptive of the sharpness and contrast of the luminance (black and white) component of the image, than the chrominance (colored) components. Thus, in any astronomical image, planetary or deep-sky, it is not a disaster if the colored components are not as sharp as the black and white ones.

More Image Stacking and Resampling

We can now see the main goal of the modern planetary CCD observer when he is at the telescope, taking images: it is to take as many well-exposed, sharp, images as possible, before the planetary features have moved by, at most, half a second of arc. By stacking hundreds or thousands of images together, the resultant planetary image looks super-smooth and noise is reduced by as much as ten times when compared to a single image. We can also see why a fast image download time is so important to the planetary observer. Amateurs using sensitive video cameras and webcams have also achieved superb results in this way. Frame-grabs from camcorders, security cameras, and webcams are, typically, quite noisy when compared to single images from a cooled CCD camera, so stacking images to reduce noise is even more important in this context.

Stacking resampled images also provides a powerful trick in compensating for a small image scale. Imagine that you just cannot obtain sharp images of planets unless your exposure times are much less than 0.2 of a second. So, to get a good signal-to-noise ratio you have to reduce your image scale to about 1 arc-second per pixel. Unfortunately, at these image scales Jupiter's limb will start to look distinctly jagged even when the image is resampled; Saturn's Cassini division looks even worse, unless you move to 0.5 arc-seconds per pixel. However, when you resample (say, to a four times larger image) a dozen or more images of the planet and then stack and average them to pixel accuracy (0.25 arc-seconds with the new image size), a miracle occurs; suddenly a smoother higher-resolution image appears, as if by magic! What you have done is to achieved a higher effective sampling resolution by statistically averaging loads of enlarged undersampled images. All of the top planetary imagers use this technique and it is one of the most powerful tools the digital imaging era has provided!

When the trained human eye sees that fleeting planetary detail on nights of indifferent seeing, it is seeing it in a fraction of a second. To capture that electronically, exposures need to be very short too, necessitating smaller than desirable image scales. However, the multiple stacking and resampling technique compensates for these small image scales and makes it possible for observers with average seeing to get excellent results. The amateur astronomer Robert J. Stekelenburg has produced a useful freeware program called *AstroStack* which is very useful for stacking planetary images together and is incredibly user-friendly. It stacks FITS or BMP images. It can be found at Robert's Astrostack Web site (see the Appendix).

All of the good CCD imaging packages contain routines for stacking images, LRGB or otherwise, together. Some also have features based on the CMY technique, which is slightly better in terms of efficient use of photons. So if you have Software Bisque, SBIG, *Maxim* or *AIP* software, image stacking will not be a problem.

Perfecting Image-Processing Techniques

I would not want readers to think that a single planetary image can't produce a good picture; with careful use of image processing to reduce noise, even single images can produce good results. However, stacking multiple images is much preferred!

Something I haven't yet mentioned and, perhaps I should at this stage, is the "drizzle" technique. This is a resampling method which is used by professionals and some amateurs and relies on the telescope drive drifting by less than a pixel between exposures to be stacked. The technical term for this motion is "dither". Again, higher effective sampling is achieved by this technique by virtue of the statistical brightness information derived from multiple images. The good news is that Christian Buil's freeware *IRIS* contains a built-in Drizzle command. It also contains some powerful "wavelet" processing which is useful for getting the best out of single planetary images.

Although we have now examined the tricks used by the world's best planetary imagers, it may be instructive to list, in sequence, the steps used by non-webcam users to process their images. These are:

1. Select the best set of luminance and RGB images in a (typically) 2- to 3-minute imaging period (SBIG's *PlanetMaster* software can be useful for this phase).
2. Sharpen each image using an unsharp mask routine.
3. Resample all the images to a consistent larger size.
4. Stack all the images (to pixel accuracy) of each color together for L, R, G, and B.
5. Merge the final L,R,G, and B composites into a true color image.

The unsharp mask tool is, undoubtedly, the most powerful routine used by the planetary imager, but it needs using with care, so that artefacts are not created. The name dates from the era when photographers would create a blurred mask of an original negative in the darkroom. This mask would then be used as a filter to suppress large-scale, low-frequency changes and enhance fine high-frequency detail. Perhaps the easiest way to visualize what this can do is to remember the problems photographers had coping with Jupiter's bright zones and darkened limb. Subtle detail in the bright zones was hardly ever captured, it was just blotted out by the background, and the areas at the limb just disappeared into the night when high-contrast paper was used. An unsharp mask cures both these problems at a stroke. Unfortunately, if single images are used, an unsharp mask will accentuate any of those dust doughnuts and other artefacts that escape the flat-fields. One way of avoiding this problem is to apply a mild Gaussian blur before the unsharp mask. This will take sharp dust artefacts out.

Unlike the world's leading planetary imagers I am an occasional planetary imager, always battling with chronic seeing from a latitude of 52 degrees north! For most of my less-than-perfect planetary images I have used a single-shot color Starlight Xpress MX5c camera, often using single shots but sometimes stacking up to ten images. Many UK amateurs use the color MX5c for planetary imaging and many more simply get a single sharp monochrome image per night. For these people, here is a summary of my own processing strategy for my single MX5c images; it's the method that works best for me:

1. Contrast-stretch the raw image to fill 90 percent of the dynamic range.
2. With the unique MX5c command, synthesize color with a saturation value of 2.
3. Apply a Gaussian smoothing filter to remove single-pixel artefacts.
4. Apply an unsharp mask (radius 4, power 2 with the camera software).
5. Save the image as a TIF and quit the Starlight Xpress software.
6. In *Paintshop Pro* (or Adobe *Photoshop*) resample the image and correct for the non-square pixels, while making the image much less blocky.

7. Correct the color balance (+16 percent red and +5 percent green works for me).
8. Tweak the brightness until the brightest parts of the image are almost saturating

In this chapter I have detailed the techniques of the leading planetary imagers; undoubtedly they enjoy better seeing than most of us, but, as Damian Peach has shown, even the northerly latitudes of the UK, ravaged by jet stream turbulence and low-pressure systems, can produce brief periods of perfect seeing. If imaging fine planetary detail appeals to you, why not have a go.

CHAPTER NINE

Supernova Discoverers

Early Days

In 1985 I joined the UK Supernova Patrol. I was allocated six bright NGC galaxies to patrol, by photography. My drive could, at best, track for 30 seconds unguided; so, if I didn't guide I was limited to detecting supernovae of magnitude 14 at best. If I did guide, to reach magnitude 16 or 17, each exposure took about half an hour to set up and take, usually in the freezing cold. To take six images involved using tiny setting circles and physically pushing the heavy 36-cm Newtonian from galaxy to galaxy. Then there was the tedium of guiding on a nearby star for each exposure. After several hours of sheer human misery, the hassle was far from over. The film had to be loaded into a developing tank, developed, fixed and dried, and then checked with a magnifier. The statistical chance of discovering a supernova, from a set of six galaxies, was virtually nil; we were all flogging the same dead horse! Only the Reverend Robert Evans, in Australia, was consistently discovering supernovae with amateur equipment at that time, and he was doing it visually.

During the 1990s a revolution in supernova patrolling took place; there were a number of factors that made this possible. Firstly, CCD detectors replaced film, allowing stars of magnitude 17 or 18 to be imaged in only 60 seconds, with the most sensitive cameras (using 25–36 cm apertures). With periodic error correction on Schmidt–Cassegrains, unguided exposures of 30–60 seconds became routine. At the same time, the Meade LX200 enabled amateurs to slew from target to target within 5 arc-minutes; just enough to always get the patrol galaxy on the CCD chip. This made another breakthrough possible: amateurs could now stay indoors in the warm, while their telescope and camera did the business in the freezing cold. This latter point was a vital step, as it was often the bitterly cold, damp nights that sapped even the keenest observer's willpower.

The Paramount GT1100 Changes the Game

The US supernova patroller Michael Schwartz, director of his private "Tenagra" Observatory in Oregon, showed the way by acquiring one of the first Paramount GT1100s and coupling it to a Celestron 14 Optical Tube Assembly and back-illuminated CCD (see Figure 9.1). This arrangement could get down to magnitude 19 in 60-second exposures and it didn't break down. The Paramount, C14, and super-sensitive CCD soon became the standard supernova patroller's kit; at a combined price of $20,000 this was becoming a game few amateurs could take part in, but it was interesting to watch the competition between the supernova patrollers! In passing, it is worth mentioning that Michael Schwartz now has second, third, and fourth major telescopes. The massive Tenagra II scope is an

Figure 9.1. The original supernova patrol telescope of Michael Schwartz: a Paramount GT1100 plus Celestron 14 at Tenagra Observatory, Oregon. Photo: courtesy Michael Schwartz.

Figure 9.2. Michael Schwartz's 32-inch (81-cm), f/7 Ritchey–Chretien in Arizona. Photo: courtesy Michael Schwartz.

81-cm (32-inch) aperture, f/7 Ritchey–Chretien reflector used for supernova photometry and patrolling faint galaxies from Arizona (see Figure 9.2). This instrument reaches just short of magnitude 22 in 5 minute unguided exposures.

The replacement third scope, nearing completion as this was being written, is a 0.6-m, f/10 Ritchey–Chretien reflector, intended for deep patrol work from Arizona. The fourth instrument is another automated C14, used mainly for minor planet/comet imaging from its Arizona site.

Michael's US rival in the supernova discovery stakes, Tim Puckett, also uses a Paramount/C14/Apogee CCD combination as well as an automated home-made 60-cm reflector (see Figure 9.3). He is the world's most successful amateur supernova hunter and his observatory is located in the North Georgia mountains, USA.

Figure 9.3. The world's number-one supernova patroller, Tim Puckett, and his 24-inch (60-cm), f/8 Ritchey–Chretien in the Georgia Mountains. Photo: courtesy Tim Puckett.

The Katzman Automatic Imaging Telescope

Unfortunately amateur patrollers have serious competition from the professionals, who also have powerful supernova patrol telescopes. The most productive is the KAIT, or Katzman Automatic Imaging Telescope. This instrument is located at Lick Observatory on the top of Mount Hamilton (altitude 1300 meters), just east of San Jose, California. The KAIT project is led by Professor Alex Filippenko of the University of California at Berkeley Astronomy Department. Dr. Richard Treffers and Professor Michael Richmond were also key players in the development of the instrument. Major funding for KAIT was provided to Professor Filippenko by the US National Science Foundation, AutoScope Corporation, Photometrics Ltd., Sun Microsystems, Hewlett–Packard Company, Lick Observatory, and the University of California at Berkeley. Completion of the telescope was made possible by a generous donation from the Sylvia and Jim Katzman Foundation. The predecessor of this project was called the Leuschner Observatory Supernova Search (LOSS) which became the Lick Observatory Supernova Search (also LOSS).

KAIT is a 0.76-m (30-inch), f/8.2 Ritchey–Chretien telescope with an f/2.5 primary mirror (see Figure 9.4). It takes 80–90 galaxy images per hour and reaches magnitude 19.5 in 30-second exposures. The larger aperture enables it to reach about a magnitude deeper than amateurs patrolling with Celestron 14s and it has a focal length 60 percent longer than a C14 operating at f/11. With its Apogee AP7 CCD camera (512 × 512, 24-micron pixels) each image covers a field of view of 6.8 arc-minutes at an image scale of 0.8 arc-seconds per pixel. It has a database of over 5000 galaxies (similar to dedicated amateur patrols) and, when imaging over 1000 galaxies per night it only takes a few days to cover all the galaxies available at any particular time of the year. The KAIT team has recently started collaborating with the aforementioned Michael Schwartz, so that patrolling isn't duplicated. The joint patrol is called LOTOSS (Lick Observatory and Tenagra Observatory Supernova Searches). In 1998, 1999, 2000, and 2001, KAIT discovered 20, 40, 38, and 68 supernovae respectively. It also discovered

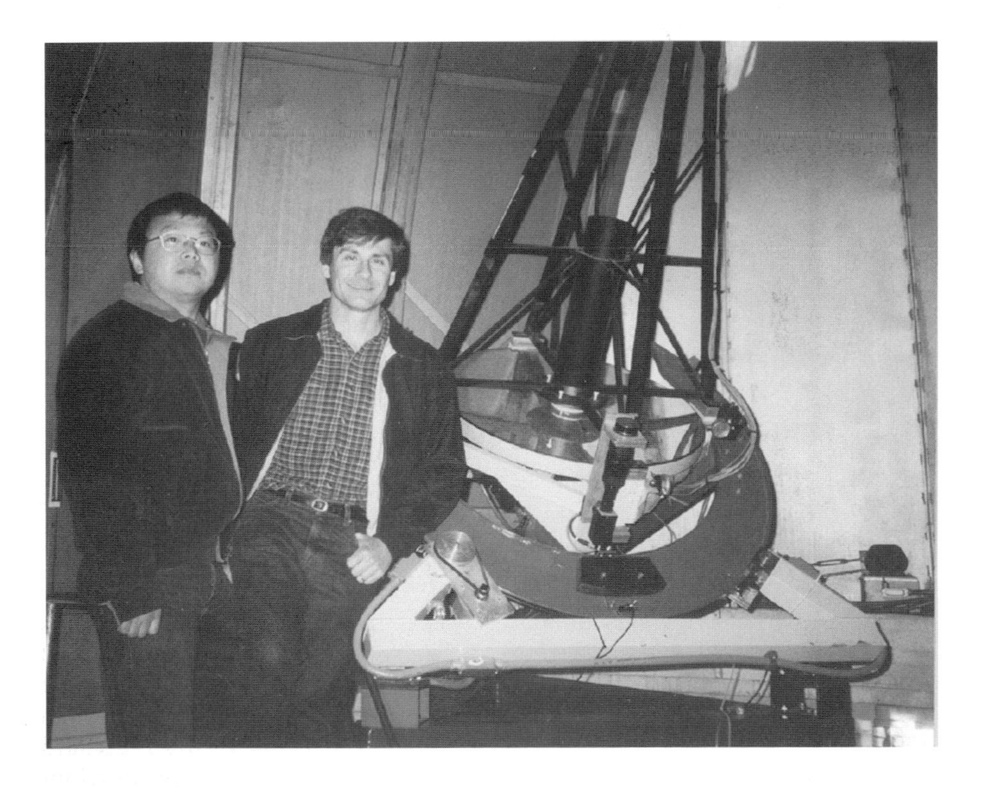

Figure 9.4. The world's most productive professional supernova patrolling telescope, the KAIT (Katzman Automatic Imaging Telescope) and principal investigators Weidong Li and Alex Filippenko. Photo: courtesy Weidong Li/Lick Observatory Supernova Search.

19 novae during this period, mainly in the Andromeda Galaxy, M31, and its satellite galaxies M32 and M110, and two comets.

As part of LOTOSS, Michael Schwartz discovered five supernovae with the original Tenagra III telescope in 2000 and eight supernovae with the same instrument in 2001. Two of Michael's discoveries are shown in Figures 9.5 and 9.6.

KAIT and other professional patrols use automatic software checking to detect supernovae. These systems align each galaxy patrol master image to the new patrol image and then subtract the master from the new image. Anything left over might be a supernova (or not) and is flagged up by the software. This might seem like an impossible system to beat, especially if the skies above the professional site are rarely cloudy. However, things are never as simple as that. As any supernova patroller will tell you, images of the same galaxy never look the same from night to night, due to changes in the altitude of the field, sky transparency, focus changes, and other factors. On an exceptional night a patrol image can sometimes go deeper than a quality master image, producing several possible supernova suspects. Professional checking methods are far from perfect and even KAIT can't cover the whole sky every night, so the amateurs still have gaps to exploit. KAIT has a few specific sky restrictions too, due to the telescope dome: it can't patrol

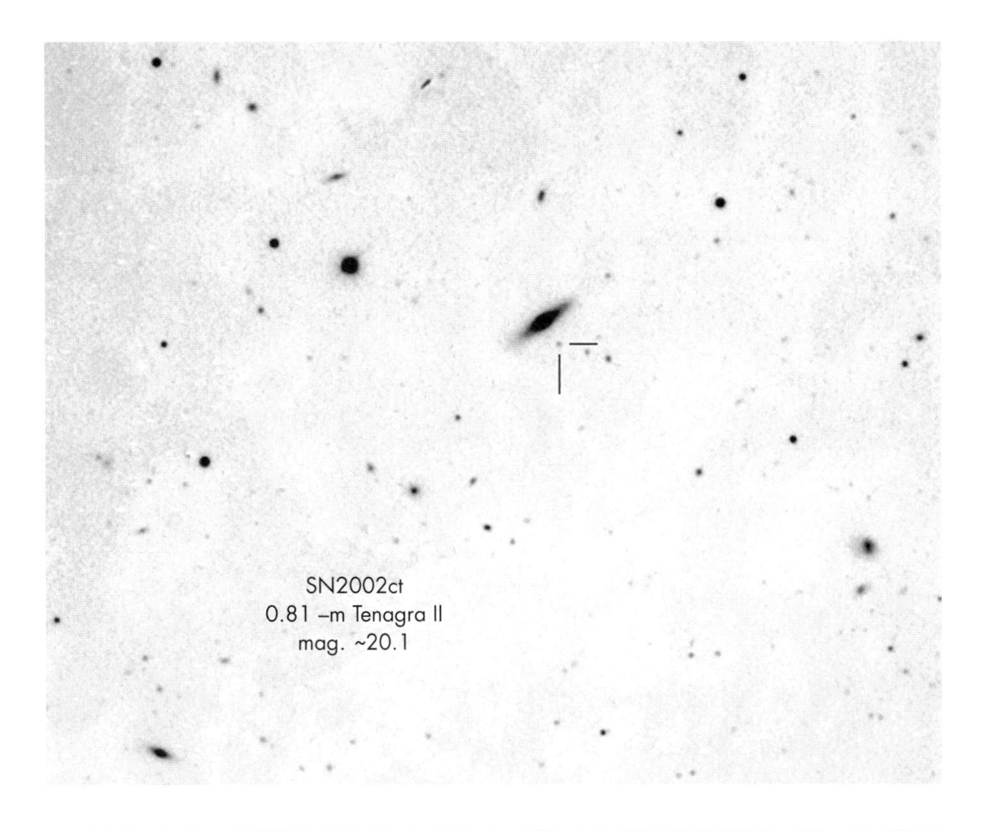

Figure 9.5. SN 2002ct; at mag 20.1 the faintest ever amateur supernova discovery, by Michael Schwartz, with his 0.81-m Tenagra II Ritchey–Chretien. Photo: courtesy Michael Schwartz.

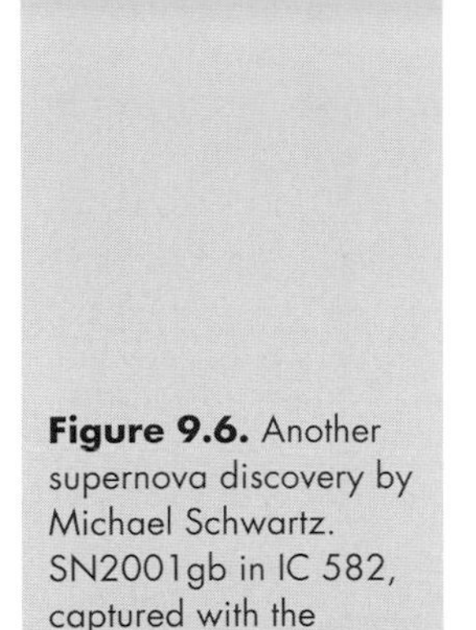
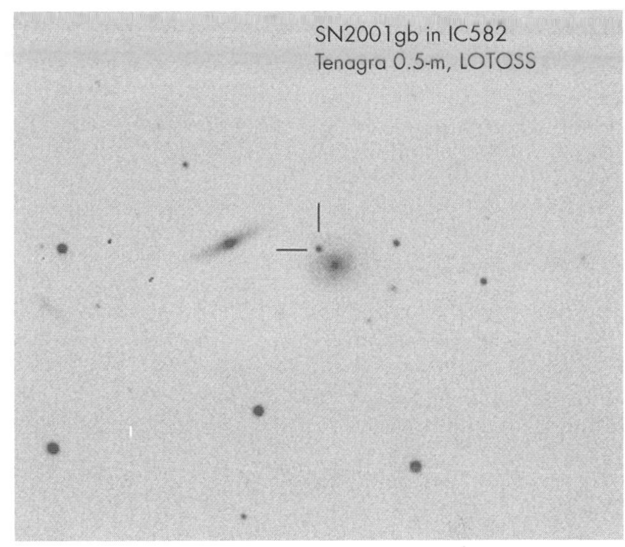

Figure 9.6. Another supernova discovery by Michael Schwartz. SN2001gb in IC 582, captured with the Tenagra 0.5-m. Photo: courtesy Michael Schwartz.

above +70 degrees Dec (or below −34 degrees Dec). It also can't patrol regions low in the east or west if they are more than 73 degrees from the south meridian. Also, with so many galaxies available and supernovae able to occur on any night, there are always plenty of galaxies available for everyone. The world's leading amateur supernova hunters have databases and master images of up to 10,000 galaxies stored on their hard disks.

Avoiding False Alarms

Discovering real supernovae – as opposed to generating false alarms – is something that should be carried out in a professional way. If a new object appears on a single-galaxy patrol image it could be due to a cosmic ray strike on the CCD, a glitch in the camera hardware/software, an asteroid, a comet, a variable star, or a supernova already discovered by others. In every case of a suspect supernova a second image must be acquired as soon as possible and, if cloud and darkness permit, a whole series of images over several hours need to be secured to check for motion of the object. In the past, the biggest source of amateur false alarms has been when galaxies are imaged for the first time with a new CCD camera or telescope and compared to images taken with different equipment; it is imperative to compare like with like and patrol images must not be checked against weaker master images. Experience in astrometric measurement is also required: a suspect's position needs to be measured using one of the numerous astrometry packages available. Planetarium packages like *Guide 8* or *The Sky* should also be on hand to check whether comets, asteroids, or variable stars are in the area. The Internet is also a vital resource for checking suspects; in particular, the CBAT/Minor Planet Center has a supernova suspect check page (see the Appendix). This Web facility will tell you in seconds if there is an asteroid

Figure 9.7. Guy Hurst, editor of *The Astronomer* magazine, in the TA Headquarters, October 2002. Photo: courtesy Guy Hurst.

anywhere near your chosen galaxy. A copy of the *Real Sky* Digitized Palomar Sky Survey (DSS) is also a huge advantage here and individual frames can also be downloaded from the DSS Web page (see the Appendix). The limiting magnitude of the Digitized Sky Survey is fainter than mag 19 and the depth and resolution is very similar to that of amateur supernova images. However, the most powerful method for checking a possible supernova discovery is consulting experienced fellow amateur discoverers.

One of the most experienced amateur nova/supernova patrol groups is *The Astronomer* team, organized, since 1974, by Guy Hurst from the *The Astronomer* Headquarters in Basingstoke (see Figure 9.7). The UK nova/supernova patrol is co-ordinated by Guy and has been responsible for verifying over 80 supernova discoveries. *The Astronomer* has an excellent Web site (see the Appendix) and Guy Hurst can easily be contacted via that site (or see the Appendix). The UK has three leading supernova discoverers, namely: Mark Armstrong, Tom Boles, and Ron Arbour. They are pictured with their equipment in Figures 9.8, 9.9, and 9.10. In late 2003, Mark, Tom, and Ron had discovered 55, 60, and eleven supernovae respectively with Mark and Tom averaging one discovery per 3,000 patrol images during 2003.

a

Figure 9.8. a Mark Armstrong, the world's number-three supernova patroller, at his UK observatory control center. The computers control a Celestron 14/Paramount GT1100 and a 30-cm Meade LX200 as well as the corresponding CCD cameras. Two more Paramounts were being planned at the time of writing! Photo: courtesy of Mark and Claire Armstrong.

(Figure 9.8. b, see overleaf)

Mark Armstrong's Patrol Statistics

Let's now have a look at the strategy of a leading supernova patroller, Mark Armstrong. No supernova had ever been discovered from the UK until 23 October 1996 when Mark discovered his first in the Galaxy NGC 673. It was magnitude 16.5, designated 1996bo and co-discovered by the Beijing Astronomical Observatory in China. Mark discovered 1996bo using a 25-cm Meade LX200 and a Starlight Xpress CCD camera.

Less than six months later, Stephen Laurie, another UK patroller, discovered the second UK supernova, also using a 25-cm LX200, but with an SBIG ST7 CCD. Tom Boles' first discovery came six months after that and then things just got silly with over 70 further discoveries in the next six years! The astrophotographer and astro-imager Ron Arbour had been the first dedicated UK supernova hunter. He started searching photographically around 1980 and built a computer-controlled 16-inch Newtonian with which he searched the skies until 1997, when he switched to a 30-cm LX200 (which he has modified extensively). In 1998 Ron discovered his first supernova; the switch to the ubiquitous Meade SCT delivering the goods within the first year!

Although Ron was not the first UK supernova discoverer, it was his quest that originally inspired Mark Armstrong, and Mark's first discovery that inspired

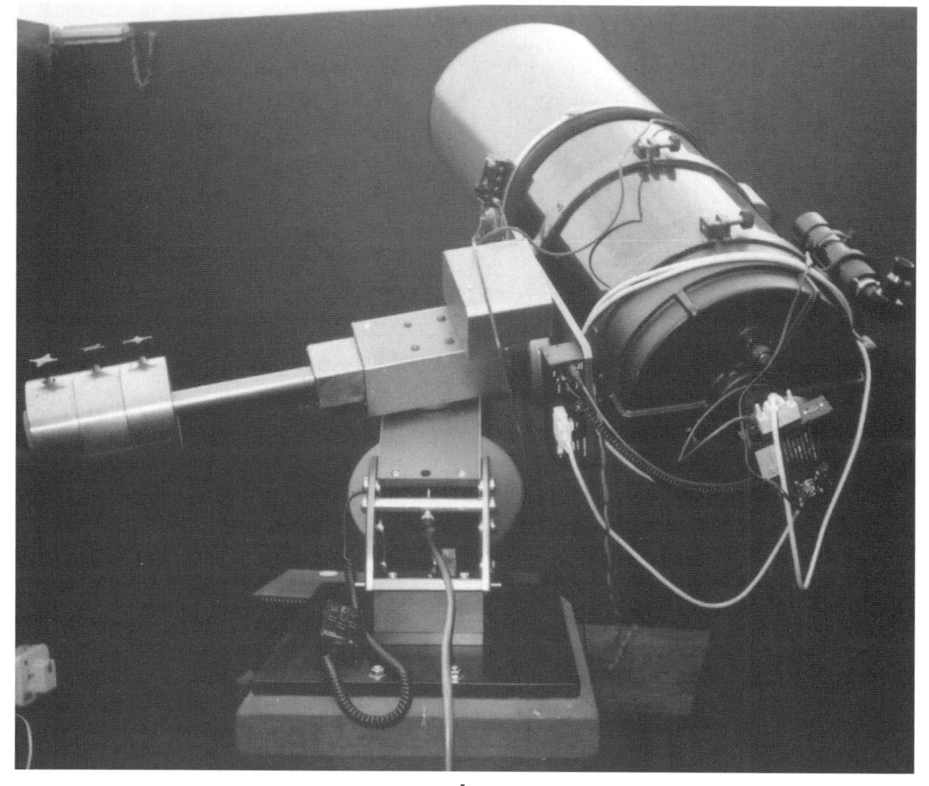

b

Figure 9.8. b Mark Armstrong's Paramount GT1100 with Celestron 14, SBIG ST9E CCD, and heated dew cap. This instrument alone has discovered more than 20 supernovae. Photo: courtesy of Mark and Claire Armstrong.

everyone else. Mark started patrolling on 21 June 1995 with the aforementioned 25-cm LX200 and Starlight Xpress SX CCD. With this system he carried out 45,585 galaxy patrols and discovered four supernovae. He then switched to a 30-cm LX200 plus Hale (back-illuminated CCD) camera and carried out 73,716 patrols making 11 discoveries. With the same LX200 and an Apogee AP7p camera he netted four discoveries from 30,657 patrols. Finally, with his most efficient system, a Celestron 14/Paramount/ST9 CCD combination he netted 16 discoveries from a mere 44,652 patrols, up to early 2002! Three UK supernovae are shown in Figures 9.11, 9.12, and 9.13.

When Mark's average discovery statistics over the seven-year period from 1995 to early 2002 are calculated, the effort involved in checking over 200,000 galaxy images are revealed. In that period Mark has:

- Patrolled the sky (often through cloud gaps) every 2.8 nights;
- Spent 4.4 hours per clear/partly clear night patrolling;
- Averaged 223 galaxies per night;

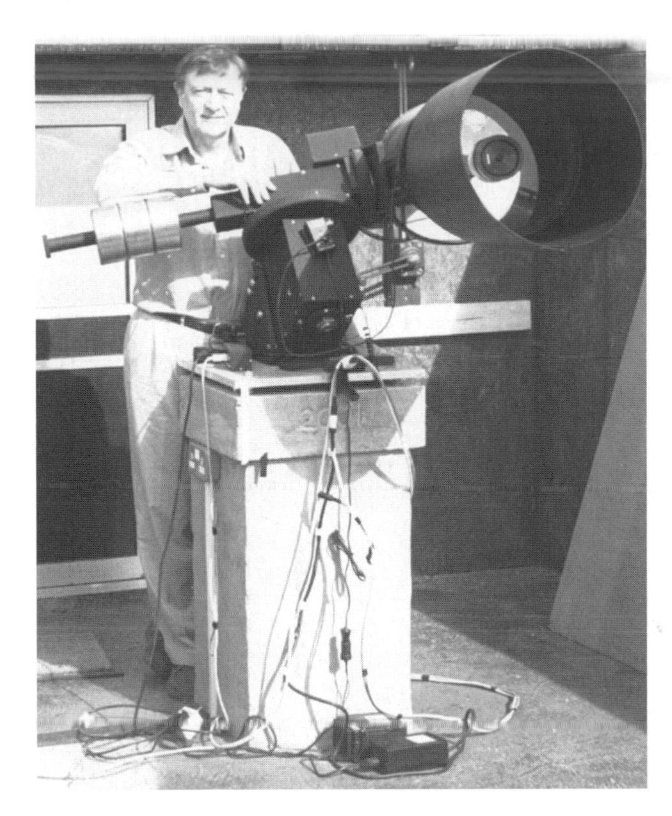

Figure 9.9. Tom Boles, the world's number-four amateur supernova CCD patroller, with his Celestron 14 and Paramount GT1100s at Coddenham, Suffolk, UK. Photo: courtesy of Nick James.

- Averaged 50 galaxy patrols per hour;
- Discovered one supernova per 5260 galaxies checked;
- Discovered one supernova every 24 clear/partly clear nights;
- Discovered one supernova for every 103 hours of patrolling.

I think these statistics show that even if you have the right equipment, you still need a huge amount of dedication, perseverance, and spare time to discover supernovae.

At the time of writing, only Tim Puckett and Michael Schwartz had discovered more supernovae than Mark Armstrong and Tom Boles, but, for me, Mark and Tom's discoveries are even more impressive: they patrol mainly through gaps

Figure 9.10. a Ron Arbour's observatory at South Wonston, equipped with a 0.3-m LX200: viewed from the northwest. **b** Ron Arbour's observatory at South Wonston, viewed from the west. Photos: courtesy Brian Knight and Ron Arbour.

between the weather systems that batter the UK and they both patrol alone, without co-investigators. Of course, all these achievements are dwarfed by Bob Evans of Australia who made all his 38 discoveries visually, most before the era of CCD patrolling!

Figure 9.11. Supernova 2001ib in NGC 7242, co-discovered by Ron Arbour and Mark Armstrong. Imaged here by the author, with his 0.3-m LX200 and SBIG ST7 on 14 December 2001. Photo: copyright Martin Mobberley.

Strategies

As well as dedication, the amateur supernova discoverers have developed strategies which make their patrol work efficient and maximize the chances of discovery. With so much competition out there this is essential. The standard Celestron 14- or 12-inch LX200 instruments have very long focal lengths which are rarely tamed with telecompressors by the supernova patrollers. At first sight this seems strange, as having a wider field of view in areas like the Virgo cluster might enable two or three galaxies to be captured in one exposure. However, the successful discoverers have found that an image scale of between 1 and 1.5 arcseconds per pixel (almost planetary imaging scales) give that extra bit of image scale necessary to separate faint supernovae from the bulge of the parent galaxy.

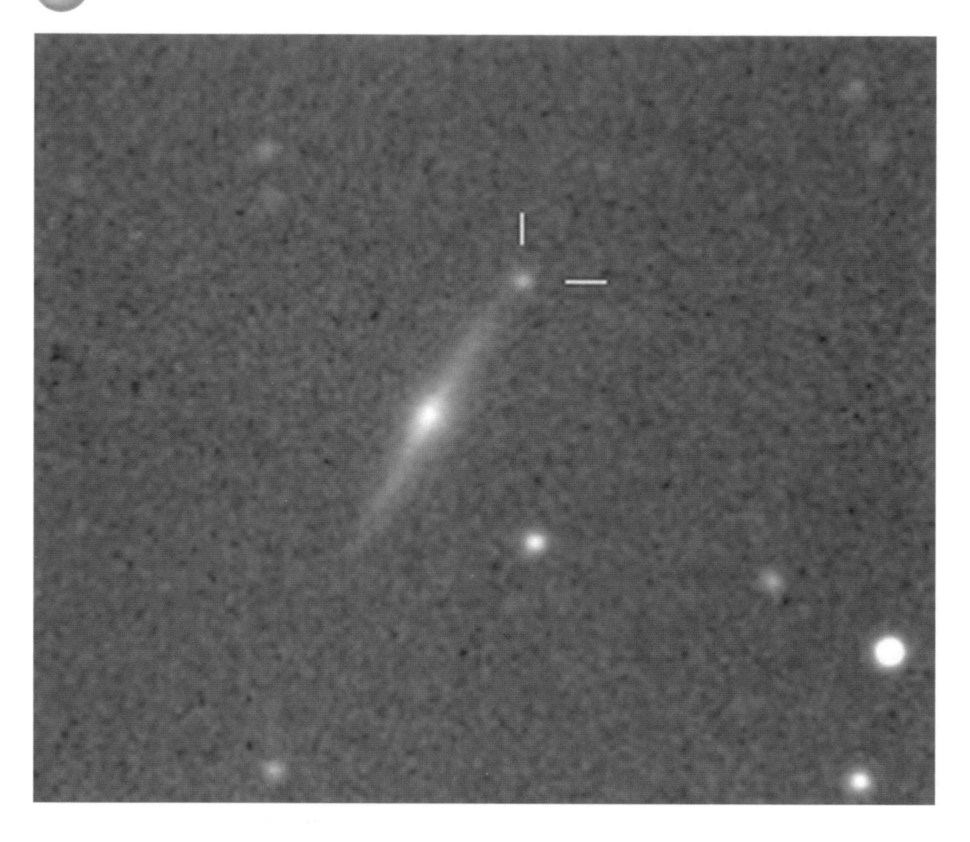

Figure 9.12. Supernova 2002bx in IC 2461, discovered by Tom Boles. Imaged here by the author, with his 0.3-m LX200 and SBIG ST7 on 5 April 2002. Photo: copyright Martin Mobberley.

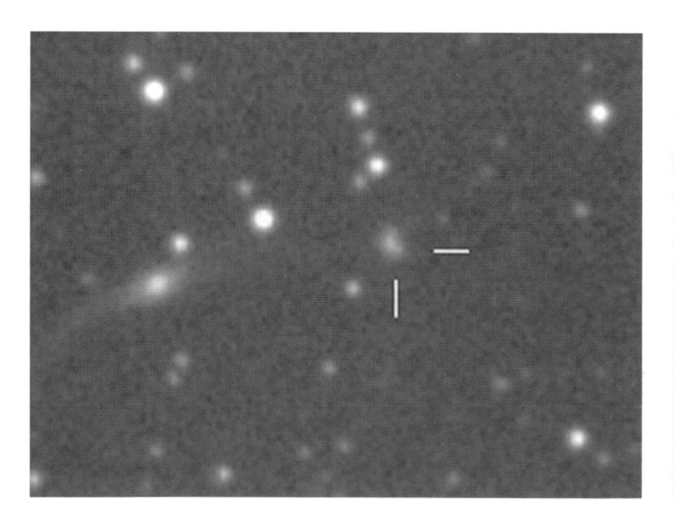

Figure 9.13. Supernova 2001iq in UGC 12032, discovered by Mark Armstrong. Imaged here by the author, with his 0.3-m LX200 and SBIG ST7 on 1 January 2002. Photo: copyright Martin Mobberley.

It also reduces the sky brightness, per pixel. Even if the local seeing would seem not to justify such image scales, so many 16th and 17th magnitude supernovae are discovered close to the galaxy core that the extra scale seems to make them that little bit more detectable. Despite the fact that such long focal lengths reduce the CCD field to under 10 arc-minutes (in most cases) it is still more productive to try to get more than one galaxy on the chip. In the vast majority of cases this is not possible, but in tightly packed clusters it can be done, meaning you can patrol two or even three galaxies per image. On this subject, The Webb Society recently published an excellent issue of their *Quarterly Journal* which may be of great interest to the supernova patroller. Issue 122 (Autumn 2000), compiled by Miles Paul, was entitled *An Atlas of Galaxy Trios* and consisted of 137 images of galaxy triplets from −32 degrees to +90 degrees declination. The selection criteria was that all the galaxies were brighter than mag 16.5 and their separation was within 10 arc-minutes. Even when no galaxy doubles or trios can be fitted on the chip, it is still far better to patrol galaxy clusters than single galaxies.

Despite the slewing abilities of modern telescopes, slewing half a degree to the next galaxy is far quicker than slewing half way across the sky! Remember, patrollers are trying to match KAIT's capture rate of 80 or 90 galaxies per hour. Some supernova patrollers start their nightly patrols with the most convenient cluster north of +70 degrees Dec (the regions where KAIT can't go). Other diehards concentrate on regions in the near-dawn sky (where fewer amateurs patrol). With CCDs the Moon should not be considered a barrier to observing. Photographic and visual patrollers used to have a week's break between waxing and waning Moon phases, but short CCD exposures with f/10 and f/11 instruments are pretty well immune to interference from moonlight. OK, at full moon you can't patrol in the same part of the sky as the Moon, but you can easily image 16th mag stars in any other part of the sky.

Lest it be thought that discovering supernovae is just a competition between suitably equipped amateurs, it should be emphasized that discovering supernovae is vital to our increased understanding of the Universe. Professional astronomers are especially interested in observing Type Ia supernovae as they all peak in brightness at about the same absolute magnitude. This makes them useful as standard candles; i.e., when a Type Ia supernova has its brightness measured and its spectrum recorded you can get a pretty good idea of how far away it is and how fast the galaxy it resides in is receding from us. This helps refine our understanding of the expansion of the Universe – powerful stuff!

Supernova patrolling is not for everyone, but if you have the time, the dedication, and the equipment, it is one of the most prestigious and rewarding areas of modern amateur astronomy. If you have good equipment, but not the time to patrol and check hundreds of galaxy images per night, the next chapter may be of more interest to you!

CHAPTER TEN

Deep-Sky Perfectionists

When CCD cameras first appeared on the amateur astronomy scene they were extensively used for taking monochrome images of 60 seconds duration, to capture star and galaxy fields as deep as 10 or 20 minute exposures would reach on photographic film, i.e., down to mag 17 at best. At last, there was no need to guide an exposure, as long as your drive could occasionally track for 60 seconds! Needless to say, amateurs soon got used to the new technology and wanted to get more out of it. Just how deep could you get with really sensitive CCDs? What if you used longer focal lengths and very long exposures from a dark site? What if you used a second CCD to guide the telescope? And what would a deep tri-color amateur CCD image look like?

With all deep-sky images, photographic or CCD, there are only a few factors having a bearing on the quality of the images obtained. The four most important factors are:

- Accuracy of focusing;
- Accuracy of guiding;
- Length of exposure (signal-to-noise);
- Freedom from light pollution (signal-to-noise).

It is remarkable how often the first two basic factors are neglected – this always leads to inferior results. Light pollution can only be avoided by observing from a dark site; the other three factors are much easier to deal with!

Polar alignment is also an issue which is sometimes neglected despite all the information available on polar alignment methods. This might seem irrelevant at first, especially if images are being guided; however, the guide star is invariably some distance from the center of the field and any polar misalignment will cause the imaged field to rotate about the guide star. Owners of the Paramount ME with its *TPoint* system can easily polar-align to within a few arc-seconds in a few nights of trials; owners of other modern SCT instruments should find aligning within arc-minutes of the pole a trivial exercise when using the instructions in the telescope manual. If your long exposures do suffer from field rotation, you can partially recover the situation by taking short exposures and aligning/stacking them with a two star alignment routine (such as found in *Maxim DL* or *AIP*). However, precise polar alignment is the only way to get the best results.

"Darks" and "Flats"

It should also go without saying that, for the very best results, dark frames and flat-fields *must* be taken for every high-quality deep-sky exposure. The dark frame is simply an exposure of matching duration, with the telescope capped, for every image taken.

This dark image is a record of *only* the thermal noise generated during the long exposure. It is then subtracted from the raw image, which leaves the sky background as the only major source of noise.

Flat-fields were discussed in Chapter 8. The easiest way to get a flat-field is to image the twilight sky with the telescope and CCD camera, as soon as twilight is dark enough to be imaged. The exposure should be set so the CCD image half-saturates, i.e., the picture is a light gray. The resulting image is then a record of the vignetting of the telescope's optics, drawtube, and any dust specks on the CCD window. The main image can then be *divided* by the flat-field – and all the optical imperfections magically disappear!

The necessary software supplied with modern CCD cameras leads the user through these processes.

Focusing

I have already talked about focusing in the context of planetary imaging. For deep-sky imaging the requirements are less critical, but no less attention should be paid to the subject. Once again, a bright star can be used to attain critical focus. If exposures of a few seconds are used, the *diffraction focusing technique* can be used. In this method, a star's diffraction spikes are used as an indicator of sharpness of focus. If you have a Newtonian reflector with a thin four-vane spider (secondary mirror holder) a bright star will end up looking like the traditional Christmas card "Star of Bethlehem" with four spikes. If the star is even slightly out of focus each spike will look double: this is a useful guide for focusing.

If you have a Schmidt–Cassegrain, you can easily make an artificial set of diffraction-creating vanes and place them over the corrector plate for focusing.

Another approach is to make a *Hartmann mask*. In this approach, you need a full-aperture cap which fits over the end of the telescope tube/dew cap. A series of four circular holes about 50 mm in diameter are cut out of the mask at the edge of the telescope aperture and at 90-degree intervals. Because each of these holes acts as a small telescope with a resolution of about 2 arc-seconds, if the telescope is out of focus the CCD image will show *four* images. These slowly merge into one as the image comes into focus.

One thing that you need to bear in mind when taking long-exposure images throughout the night is that the focus point will usually change appreciably as the telescope cools down. For high-quality work, planetary or deep-sky, checking the focus regularly is essential.

I have previously mentioned that Jim's Mobile Inc. (JMI) sell their "New Generation Focusers" with optional digital read-out and motor focusing to make the job easier; they also market an (expensive!) temperature-compensating focuser which remembers the focus adjustments necessary at each temperature and applies them automatically. Michael Schwartz uses this system when supernova patrolling with his Celestron 14.

Guiding

Today's deep-sky imager has a number of options available for keeping a guide star stationary in the telescope field and subsequently delivering a quality image. Guiding visually with a guiding eyepiece is still a possibility, but is rarely used by the modern amateur.

In reality the choices available are:

1. Using a dual-chip autoguiding CCD camera from SBIG;
2. Using a separate CCD autoguider attached to a separate guidescope or attached to an off-axis guider;
3. Using the Starlight Xpress on-axis *Star 2000* autoguiding system;
4. Using SBIG's cameras with their AO-7 adaptive optics unit.

I have already described the products above in other chapters, so I won't repeat the technical details. Most of the world's leading amateur deep-sky imagers use autoguiding CCD cameras from SBIG with or without the AO-7 Adaptive Optics unit. The AO-7's main advantage is that it moves its lightweight mirror to autoguide short distances, as opposed to instructing the drive to move the whole inertial mass of the telescope. In practice, both systems, when used with a good drive, can detect subpixel drifts of the guide star and correct the guiding within a fraction of a second. If the atmospheric seeing is exceptionally good and long focal lengths are being used (1 arc-second per pixel and finer image scales) the AO-7 can make a detectable difference to an image, but it's not essential.

Length of Exposure

Most amateur images of deep-sky objects are short exposures, typically of 1 or 2 minutes duration, or maybe several 1-minute exposures, stacked together.

When compared to really long (single or stacked) exposures they look noisy and grainy, with a gritty, gray, pixellated background. They rarely have the quality or smoothness that a long exposure produces. This is all a matter of signal-to-noise ratio. In a short exposure the faint details in the outer arms of galaxies, or the edges of nebulae, are barely visible above the sky background brightness. The standard trick to compensate for the details being faint is to carry out a contrast stretch, but while this may make the features brighter it also boosts the noise.

Long exposures are always necessary to get the best results; today's leading imagers are typically using very long exposures that are reminiscent of those used by amateurs using hypersensitized Kodak 2415 film in the 1980s – but the results obtained by these amateurs are nothing short of fantastic: sharp, smooth, tri-color images down to magnitude 22.

LRGB

Up to a few years ago most deep-sky imagers took three images in red, green, and blue, stacked them up, and that was the final image. But there was still considerable room for improvement. Red, green, and blue filters attenuate the light hitting the CCD pixels by a considerable amount in order to get a true color image.

What the most advanced amateurs wanted was a quality image with the smoothness and magnitude limit of an unfiltered image, plus the color of a filtered tri-color image. Thus the concept of the LRGB (L = luminance) deep-sky image was born. Because of the fact that the human eye–brain combination is not especially sensitive to color resolution, the filtered images did not even need to be taken at the same resolution as the monochrome image. The filtered exposures could be binned 2×2 for greater pixel sensitivity, or even taken with a different telescope! As long as the images were scaled and stacked up correctly, the color resolution was not critical. The big advantage of an LRGB image is that it is, effectively, a high signal-to-noise luminance image retaining the color, but not the noise, of an RGB image. Modern image-processing packages can stack LRGB images or the stacking can be done in a package such as Adobe *Photoshop*. The sequence of events in the build up of a typical LRGB deep-sky image is not that dissimilar to compiling a planetary image, except that the exposures are much longer and can be made over many nights, not 2–3 minutes! The procedure is as follows:

- Expose a long monochrome luminance image(s).
- Expose the filtered RGB images.
- Apply the relevant dark frames/flat-fields to the obtained images.
- Use an appropriate package to add and align the RGB color information to the deep, unfiltered luminance image.

A few extra points are worthy of mention. First, there is no reason why multiple filtered or monochrome images should not be aligned or stacked together to form each individual L, R, G, and B image. Second, the duration of the R, G, and B images should be matched to the sensitivity of the CCD camera/filter combina-

tion used. Older CCD cameras were far more red-sensitive than blue-sensitive. Typical values for the best LRGB exposures might be 60 minutes L, 20 minutes R, 20 minutes G, and 30 minutes B, with filtered images typically binned 2 × 2 as sensitivity, not resolution is the issue where the color exposures are concerned. Third, I have not yet mentioned any advanced image-processing stage appropriate to a deep LRGB image. The best technique, which is almost universally applied by the leading amateurs these days, is known as the Digital Development Process, or DDP.

DDP

When deep-sky CCD imaging was in its infancy, the goal of image processing seemed to be to contrast-stretch everything to bring out the faintest detail. This did the job, but for objects like galaxies it always left the central cores resembling totally overexposed white blobs! What was needed was a routine that suppressed the near-saturated galaxy cores, while boosting the faint outer regions. The unsharp mask routine was sometimes used, but, by its very nature, it left the stars in the field surrounded by dark rings. Some amateurs preferred to use non-linear contrast stretches, but none of the standard ones ever seemed to get the job done correctly.

Enter Dr Kunihiko Okano of Japan (see the Appendix for his Web page URL), a Japanese amateur astronomer and fan of classic Ford Mustangs. Dr Okano developed his digital development process algorithm for processing deep-sky images so they would have a "photographic quality'. At first this might seem like a backward step, but let's not get confused; nobody wants to go back to the days before CCD detectors existed. While film is incredibly insensitive compared to a CCD, it does have a couple of very attractive features. The so-called "gamma curve" (input brightness plotted against output brightness) for film flattens off at the bright end, whereas CCDs just reach saturation and everything turns white. So, a long, deep photographic exposure of a galaxy on film will suffer less from central core white-out than a raw CCD image (of course, the CCD exposure may take a twentieth of the time to expose, but that's not what we are discussing). This means that film has the useful ability to compress the dynamic range of the galaxy's features during the exposure, so they will reproduce nicely when printed. Film has another desirable property too, in that so-called "chemical adjacency" effects add an extra degree of sharpness to edge boundaries when the film is being developed. Dr Okano virtually transformed galaxy image processing when he created his single-step DDP routine which combines gamma curve modification with a modified unsharp mask routine to bring the two most desirable qualities of photographic film into the quantum-efficient CCD world.

Combining a deep DDP luminance image with chrominance RGB data results in truly spectacular color images, especially images of galaxies. It is still possible to end up with stars surrounded by dark haloes, courtesy of the unsharp mask component, but this can pretty much be avoided if the unsharp mask radius is kept small. Image-processing programs, like *Maxim DL*, feature LRGB/DDP routines and tutorials to lead you through the process of getting spectacular images.

These techniques are powerful but will not do all the work for you. Remember that the world's best imagers always pay careful attention to the basics – focusing, guiding, long exposures, polar alignment, and, if possible, attaining a dark-sky site.

The DDP technique is the best method for producing spectacular "pretty pictures" of galaxies and nebulae but applying it does eradicate the scientific data in the image. Specifically, DDP images should never be used for estimating stellar magnitudes, either with a photometry tool or comparing star sizes. Not only does this type of routine modify the data, it modifies the sizes of the stars too. The DDP routine is very good at turning small fuzzy stars into pinpoint stars.

At this point, a brief reminder may be necessary. When purchasing a CCD camera for deep-sky imaging you may well have a choice between an antiblooming gate (ABG) or a non-antiblooming gate (NABG) CCD. As I explained in the CCD section, the ABG cameras are non-linear but prevent bright stars saturating and bleeding charge in a vertical line on the image. The NABG CCDs are more light sensitive and have a much more linear response when approaching saturation. Most amateurs will choose sensitivity and linearity and put up with the odd "bleeding" star. The choice is yours. Me, I'd recommend going for the NABG option.

Deep-Sky Imager's Telescopes: The Ritchey–Chretiens

Before we end our look at deep-sky imaging let's just have a look at what telescopes the best imagers are using. Despite the preponderance of Schmidt–Cassegrains on the market, they are rarely used by the very best deep-sky imagers. As far as I can tell, the reason for this is that the very best amateur imagers have all become total perfectionists with regard to observing site and equipment.

Fortunately, there are a few exceptions to this rule that prove that standard SCTs can be used. For example, one of the best deep-sky imagers in the UK is Gordon Rogers, who uses a 40-cm Meade LX200 plus SBIG ST8 and AO7. His telescope is spectacularly mounted in a dome atop his house (see Chapter 4)! Gordon uses his 40-cm SCT with a telecompressor giving f/6.3 and an image scale of 0.74 arc-seconds per pixel. A complete sequence of Gordon's images in the production of the final, superb, image of the galaxy M106 is shown in Figure 10.1. Further images by Gordon are shown in Figures 10.2 through 10.5. A 40-cm LX200 was also used by Adam Block, an astronomer at the Kitt Peak Visitor Center, to take stunning images. The Center's Advanced Observing Program allows amateur astronomers to hire the equipment and take advantage of the superb skies at Arizona's Kitt Peak Observatory. Adam's images are amongst the best I've ever seen and more details of the Advanced Observing Program can be found at the AOP Web page listed in the Appendix. Like Gordon Rogers, Adam's images are taken with an SBIG ST8 CCD, with 9-micron pixels, at f/6.3 and even f/10 (when seeing permits). An adaptive optics unit can also be employed where

Figure 10.1. Raw, dark, flat, luminance, and RGB images used to compile the final, excellent image of M106, taken by Gordon Rogers' 0.4-m f/6.3 Meade LX200 and SBIG ST8E/AO-7 imaging system. All images courtesy of Gordon Rogers. **a** raw; **b** dark; **c** flat
(Figure 10.1. d, e, f, see overleaf)

required. But if most of the world's leading (and wealthiest) amateurs are *not* using Schmidt–Cassegrains, what are they using?

The answer is, rather predictably, the aforementioned Ritchey–Chretien telescopes. I don't mean to be derogatory to rich amateurs. It is, in fact, the best

Figure 10.1. *Continued*. Raw, dark, flat, luminance, and RGB images used to compile the final, excellent image of M106, taken by Gordon Rogers' 0.4-m f/6.3 Meade LX200 and SBIG ST8E/AO-7 imaging system. All images courtesy of Gordon Rogers. **d** luminance; **e** red; **f** green

optical system for carrying out wide-field, high-quality imaging. The Kitt Peak AOP has just installed a 0.4-m Ritchey–Chretien by RCOS at it's refurbished visitor center (mounted on a Paramount ME) and a 0.63-m Ritchey–Chretien instrument is due for installation in 2003. What a toy to hire for a night under those dark skies!

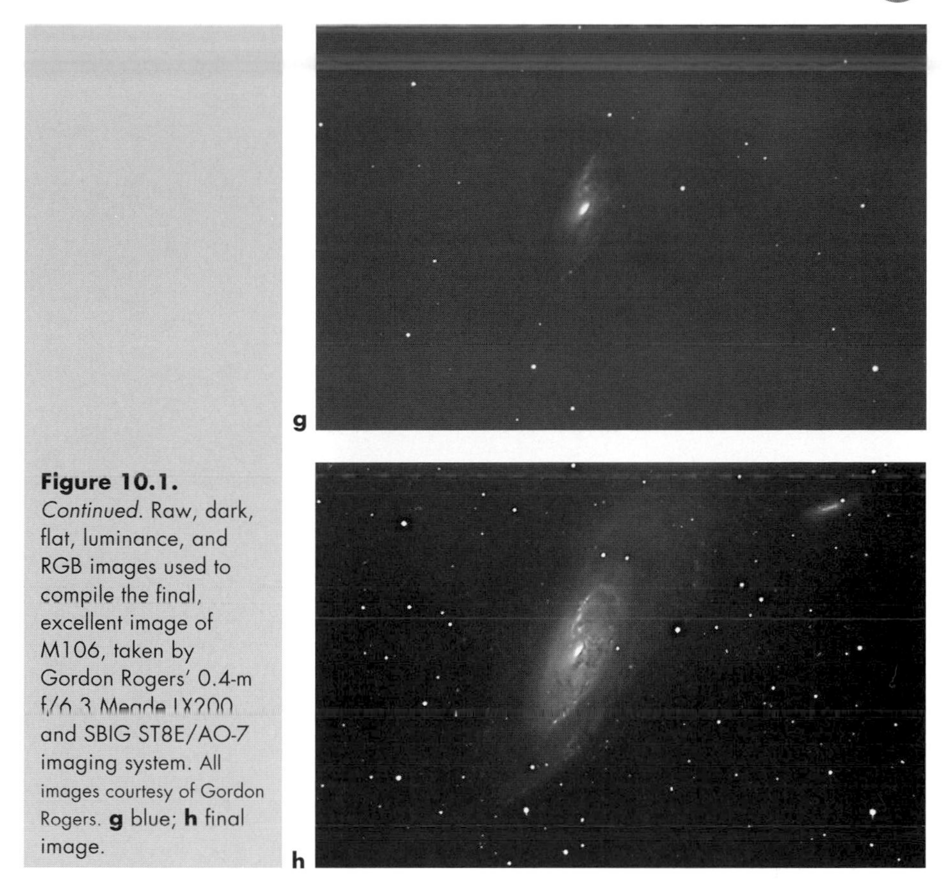

Figure 10.1. *Continued.* Raw, dark, flat, luminance, and RGB images used to compile the final, excellent image of M106, taken by Gordon Rogers' 0.4-m f/6.3 Meade LX200 and SBIG ST8E/AO-7 imaging system. All images courtesy of Gordon Rogers. **g** blue; **h** final image.

All standard reflecting telescopes suffer to varying degrees from an aberration called coma: the further away from the center of the field you go, the more the stars look like pear-shaped (or fan-shaped) blobs. With a small, f/5 (or faster) Newtonian reflector, this effect is quite apparent on stars more than 10 arc-minutes from the center of the field of view; and it's *really* apparent on a CCD image! In typical (poor) seeing conditions, the smallest star diameters on a CCD image will be about 4 arc-seconds across. So, as soon as off-axis star images are distorted by more than this amount, the effect will be very noticeable.

Schmidt–Cassegrains also suffer from coma, but Ritchey–Chretiens do not. They have complex hyperbolic mirror surfaces which produce coma-free images. They also have a couple of other advantages. There is no transparent corrector plate as in the Schmidt–Cassegrain telescope, so less light is scattered around; plus the secondary amplification factor is, typically, 2.5–3×, as opposed to 5× for an SCT, resulting in a flatter field.

The one optical *dis*advantage of a Ritchey–Chretien is that it typically has a secondary mirror 40 percent the diameter of the primary mirror, reducing contrast for planetary work. This should not be a concern for apertures above 300 mm or where a CCD camera is being used, as large apertures and image processing can compensate for a large central obstruction.

Figure 10.2. NGC 5985. 16-inch (0.4-m) LX200 at f/6.3. SBIG ST8E and AO7. Photo: courtesy Gordon Rogers.

The major disadvantage for most observers is sheer cost. Ritchey–Chretien optical tube assemblies (OTAs) are *much* more expensive than mass-produced Schmidt–Cassegrain OTAs. At the time of writing (late 2002) a 35-cm Celestron OTA costs around \$4000 and a 40-cm Meade OTA around \$8000. Corresponding Ritchey–Chretien OTAs cost \$18,000 (37 cm) and \$23,000 (40 cm) and go up to \$30,000 for the highest-specification 40-cm OTA.

We are obviously talking serious money here, and that's without a mounting! The two leading US companies competing for the Ritchey–Chretien (RC) market are Optical Guidance Systems of Huntingdon Valley, Pennsylvania, and RC Optical Systems (RCOS) of Flagstaff, Arizona. Optical Guidance Systems has been supplying 25–81 cm Ritchey–Chretien telescopes to professional and amateur observatories since 1983 and claims to be the world's number-one Ritchey–Chretien supplier. It now manufactures complete telescope systems up to 1-meter aperture and is a supplier of robotic and Internet-controlled telescopes. The company has supplied over a hundred telescopes, of various optical designs, to professional and advanced amateur observatories worldwide. RC Optical Systems (RCOS) is a newer company, but claims that it has shipped more Ritchey–Chretien telescopes worldwide, since 1998, than any other company. Figures 10.6, 10.7, and 10.8 show some of their products. RCOS specializes in manufacturing OTAs in the 25–40 cm aperture range with its most popular OTA by far being the

Figure 10.3. M82. 16-inch (0.4-m) LX200 at f/6.3. SBIG ST8E and AO7. Photo: courtesy Gordon Rogers.

Figure 10.4. NGC 4631. 16-inch (0.4-m) LX200 at f/6.3. SBIG ST8E and AO7. Photo: courtesy Gordon Rogers.

Figure 10.5. NGC 5907. 16-inch (0.4-m) LX200 at f/6.3. SBIG ST8E and AO7. Photo: courtesy Gordon Rogers.

32-cm (12.5-inch), f/9 instrument. Prices for a 25-cm, f/9 OTA start at around $12,000. For some time it has offered a 63-cm OTA as its largest aperture, but an 81-cm is promised soon.

The very best collections of amateur images I've spotted on the Internet are in the gallery sections of SBIG and RC Optical Systems (see the Appendix). I would strongly advise readers to find these pages and see what the world's best amateurs can achieve, largely with 32-cm, f/9 Ritchey–Chretiens; exquisite tri-color images to 22nd magnitude, with a smooth black background – wow!

Figure 10.6. A 10-inch (25-cm) Ritchey–Chretien Optical Systems optical tube assembly (OTA) will set you back at least $11,600. Photo: courtesy Brad Ehrhorn/RC Optical Systems.

Figure 10.7. A 16-inch (40-cm) Ritchey–Chretien Optical Systems optical tube assembly (OTA) on a Software Bisque Paramount ME mounting. This telescope is installed at the Kitt Peak Visitors Center, Arizona. The cost of OTA plus mount was roughly $40,000. Photo: courtesy Brad Ehrhorn/RC Optical Systems.

Most of these amateurs have mounted their OTAs on Astrophysics 1200GTO or Paramount mountings, to achieve a total system cost well in excess of $20,000 and often over $30,000. (Who would have dreamed, ten years ago, that there was so much money about in amateur astronomy!) Incidentally, George Willis Ritchey's optical potential was first recognized by no less a person than George Ellery Hale. The two men worked together as an awesome team at Yerkes Observatory and, from 1905, at Mt Wilson. In collaboration with the French optician, Henri Chretien, Ritchey developed the wide-field optical system that bears his name, for the 60-inch Mt Wilson reflector. However, when Hale refused to allow the design to be used on the 100-inch, the two men quarrelled, and ultimately Hale fired Ritchey.

Ironically, the Ritchey–Chretien design is now used on virtually every major professional telescope, including *Hubble*. In practice, the Ritchey–Chretien OTAs purchased by the new breed of amateur astronomers in the last five years have no overwhelming optical advantage over the Schmidt–Cassegrain where narrow-field imaging of galaxies is concerned. Coma is not objectionable over a narrow (less than 10 arc-minute) wide field and most CCDs used at the f/10 focus of an

Figure 10.8. A 25-inch (63.5-cm) Ritchey–Chretien Optical Systems optical tube assembly (OTA.) At $70,000, a professional instrument! Photo: courtesy Brad Ehrhorn/RC Optical Systems.

SCT cover a narrower field than this. There are very few galaxies bigger than 10 arc-minutes across.

Adam Block's images, taken at Kitt Peak, prove that a Ritchey–Chretien is not essential for narrow-field imaging.

However, SCTs do have a number of shortfalls, and you need to understand them before you can get great images. SCTs are notorious for suffering from "mirror flop". Their thin mirrors are supported on the central baffle tube and, as the telescope slews or tracks, the focus can subtly shift, ruining the image quality. And SCTs are also very prone to focus shifts caused by temperature change.

Both major manufacturers of Ritchey–Chretien OTAs employ tube materials that have a zero (or close to) coefficient of temperature expansion. This is achieved by using a carbon-fiber tube, or by using an aluminum/magnesium tube with invar metering rods. Temperature control systems (vibrationless fans linked to a temperature monitoring system) are available for the Ritchey–Chretien OTAs. With an SCT you will have to spend 15 minutes every time you go outside, adjusting the focus until you are happy. On many nights you'll have to adjust the focus every hour (unless you buy one of those expensive temperature-compensating focusers).

With a modern Ritchey–Chretien, you may not have to adjust the focus at all, from night to night. The quality here not only extends to the mirror support and tube materials; the optics are made from zero-expansion materials too and are hand-figured to a very high accuracy. If you do have $20,000 or $30,000 wearing a hole in your pocket and can afford the very best in optics and mountings, go for it: a Ritchey–Chretien will not disappoint!

The humble Newtonian can also be used to take superb deep-sky images. It is not prone to mirror flop and can deliver superb images, especially in longer focal ratios of f/6 or f/7. As far as I can see, the only reason they are rarely used is because of the commercial autoguider-friendly mountings of Schmidt–Cassegrains, and because the compact size of SCTs is a great attraction. Many leading amateurs – especially in the USA – haul their telescopes up mountains to get dark skies and clear air; Mt. Pinos in California is a favorite observing site for many. Lugging a 30-cm, f/7 Newtonian up a mountain is a far more formidable prospect than lugging a 30-cm LX200 up a mountain.

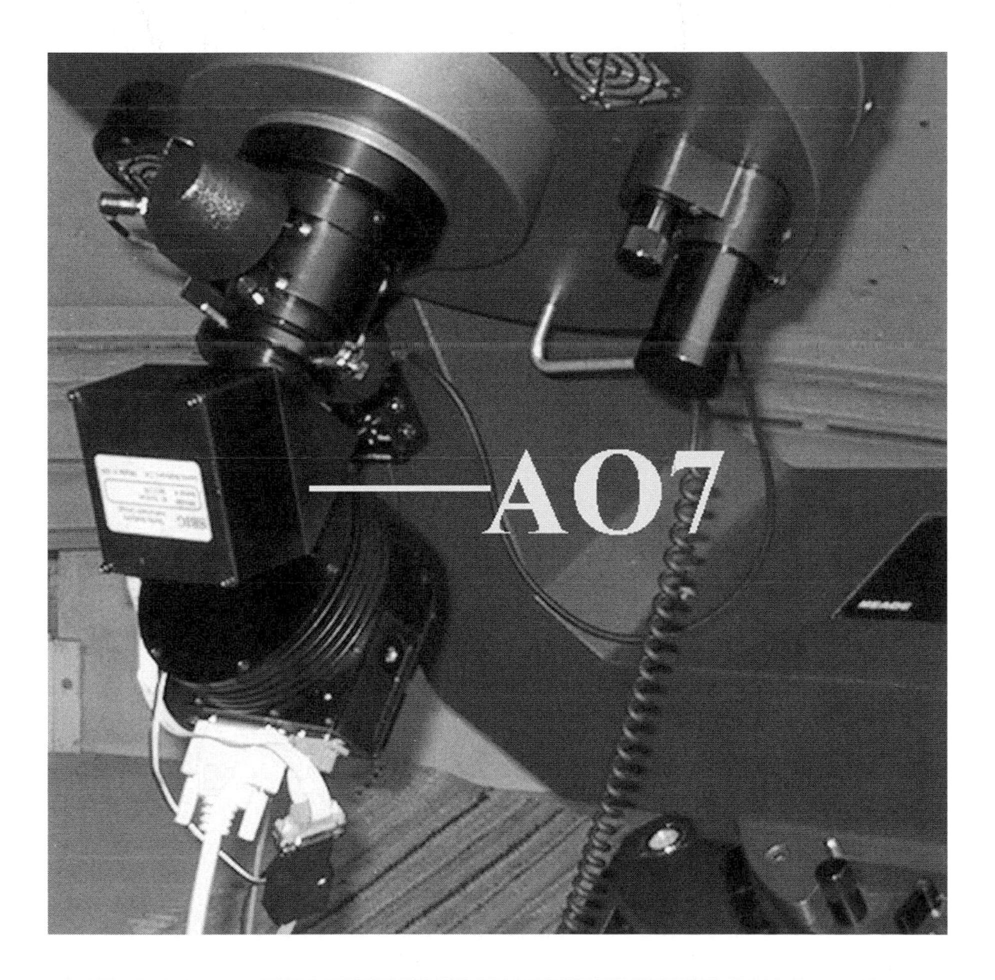

Figure 10.9. An SBIG AO7 adaptive optics unit, attached here between a CCD camera and a digital focuser, allows rapid guiding corrections without attempting to move the telescope's considerable mass. Many top deep-sky imagers use this device to deliver the best results at image scales finer than one arc-second per pixel. Photo: courtesy Arto Oksanen.

The autoguider issue is less important because "work arounds" are possible. Many commercial telescope drives have been interfaced to autoguiders and there are plenty of people on Internet discussion groups who can advise. Also, SBIG's AO7 adaptive optics unit (Figure 10.9) combined with an SBIG dual-head CCD camera, can correct for drive errors by moving its own mirror, without interfacing to the (often less-than perfect) telescope drive. The unit can cope with ±50 pixels of motion before the mirror runs out of travel, so at 1 arc-second per pixel, a Newtonian drive with a periodic error less than ± 50 arc-seconds could be used to take superb deep-sky images.

With patience and perseverance *any* good-quality telescope can be used to take spectacular deep-sky images but, if you need some inspiration, a quick look at the results of the best imagers on the Web pages of SBIG and RCOS should spur you on.

Good Luck!

CHAPTER ELEVEN

Cataclysmic-Variable Observers and Gamma Ray Burster Hunters

Outbursts of Cataclysmic Variables

Observing variable stars is something that is fascinating in itself, but has the bonus that the results are of great interest to professional astronomers. This is real science.

A category of stars that is of especial interest to professionals comprises the so-called *cataclysmic variables*. Recently, Springer–Praxis published a book which is generally regarded as the "Bible" for this type of study: *Cataclysmic Variable Stars* by Coel Hellier (ISBN 1-85233-211-5); I recommend it highly. Cataclysmic variables (CVs) are double stars, orbiting each other in such close proximity that material flows between them, producing outbursts and variability unlike any other category of variable star. They can vary in time-scales of minutes, hours, days, weeks, or months, and are favorite subjects for variable-star observers. It is not an exaggeration to say that to a keen variable-star observer, the star fields of the favorite cataclysmics can become as familiar as old friends.

Visual observers like the UK's Gary Poyner and Chris Jones have been observing CVs for years. In Gary's case, he can locate and estimate the magnitude of over one hundred variable stars per night, purely from memory! Yes, really: he has memorized the position and the comparison stars for hundreds of variables, mainly CVs (over 1000 CVs are known).

For mere mortals who don't have Gary's abilities to star hop (he uses the 45-cm Dobsonian shown in Figure 11.1) the era of the Go To telescope has opened up new possibilities. With an LX200 and a suitable software package, amateurs can

Figure 11.1. Gary Poyner and his 18-inch (45-cm), f/4.5 Obsession Dobsonian. Photo: courtesy Gary Poyner.

slew from variable-star field to variable-star field with ease and pre-program their PC with their observing schedule before it gets dark. Hazel McGee is one such amateur. She freely admits that if she didn't have an LX200 there is no way she could star-hop to all the fields, and her telescope would lie unused. Since getting a 12-inch (30-cm) LX200 (Figure 11.2) she has become one of the UK's top variable-star observers, clocking up 2500 visual magnitude estimates in 2001 alone. Hazel uses the software package *Skymap Pro* to plan her observing session and slew the LX200 from field to field during the night.

As well as observing CVs with a Go To telescope, a growing number of new amateurs are patrolling the fields of CVs with CCDs to capture new outbursts as soon as they occur. Many of these stars have their minimum brightness well below the mag 14 or 15 limits of visual observers and as there are over one thousand that are known, barely a night goes by without some outburst of a CV occurring.

In practice, amateurs tend to restrict themselves to patrolling the CVs whose outbursts recur on a reasonable time-scale and are within easy range of their equipment. As a rough guide, there are dozens of CVs under regular observation each night and about a hundred are being patrolled semiregularly for rare out-

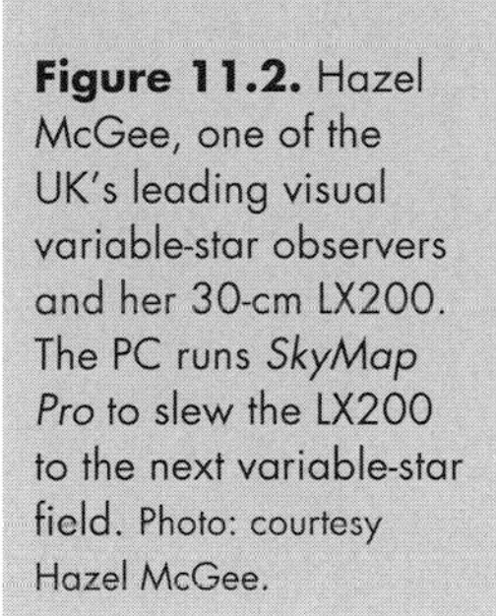

Figure 11.2. Hazel McGee, one of the UK's leading visual variable-star observers and her 30-cm LX200. The PC runs *SkyMap Pro* to slew the LX200 to the next variable-star field. Photo: courtesy Hazel McGee.

bursts. The BAA's Variable Star Section has a list of 80, rare, top priority, "Recurrent Objects" and, during the course of a year, ten to fifteen of these will typically be spotted in outburst. Once numerous outbursts are proven, stars are often removed from this list and other suspected objects added. Discovering a rare outburst of a CV is highly prized; it has happened to me once – on 20 September 2000 I took a deep CCD image of the field of V402 Andromedae (formerly 62 And) which normally sits below magnitude 20. Another CCD observer, Tonny Vanmunster in Belgium, had imaged the field two days earlier to check for an outburst, but found nothing to magnitude 17.6, but on my image it was clearly visible at magnitude 15.7. This was subsequently confirmed *visually* at magnitude 15.4 by Gary Poyner and Chris Jones. It was a big moment for me.

Fortunately I was able to acquire a chart of the region from the AAVSO Web site (see below) and had already prepared an image of the region from the Palomar Sky Survey *Real Sky* CD-ROM; otherwise, identifying the field would have been a nightmare! The potential outburst spotter needs to receive circulars from at least one of the main organizations dealing with such events to keep aware of developments. Five of the top variable-star organizations in the world are listed below. Their Web pages are listed in the Appendix. They are:

- The American Association of Variable Star Observers (AAVSO)
- The Variable Star Section of the British Astronomical Association
- *The Astronomer* magazine (TA)
- The Variable Star Network (VSNet) of Japan
- The Center for Backyard Astrophysics (CBA) (keen CCD photometry specialists)

If you want to be up to date with the latest outbursts of variable stars, VSNet will happily bombard you with apparently infinite numbers of bulletins on what is happening! If you prefer fewer e-mails, the AAVSO can keep you well informed and *The Astronomer* e-circulars always cover the rarest variable-star outbursts as soon as they occur. *The Astronomer* Web site is updated regularly with images of novae, supernovae, and variable-star outbursts and is an invaluable resource for anyone who wants to find a CCD image of a variable star in outburst.

As I've already mentioned, having a chart of the region, along with a Palomar Sky Survey image, is invaluable. It is worth building up a set of charts well before you start serious CV outburst patrolling. The AAVSO and TA are excellent sources of charts. Images from the Palomar Sky Survey can be downloaded from the Space Telescope Science Institute DSS page (see the Appendix).

Alternatively, you can purchase the highly compressed (lower-quality images) *Real Sky* CD-ROM set from the Astronomical Society of the Pacific. Another invaluable weapon in the CV armoury is the superb publication known as "Downes and Shara," more properly called *A Catalog and Atlas of Cataclysmic Variables*, also published by the Astronomical Society of the Pacific. This publication catalogues 1020 known CVs and has (mainly in the first edition) miniature, 4 arc-minute wide, images of about 500 of the CV fields. While not essential, this is a major resource for identifying objects when they are faint and when you already have wider-field charts to hand.

Variable-Star Photometry

Perhaps the most valuable contribution an amateur can make to professional astronomy (short of the discovery of a comet, nova, supernova, or CV outburst) is to plot the light curve of an unusual variable star. This area is well exploited by a growing band of new amateur astronomers equipped with CCDs. Fortunately, because there are hundreds of CVs and because amateurs are spread around the globe, some in daylight, some in darkness (and many clouded out), there are still niches to fill in this field. Traditionally, the majority of variable-star magnitude estimates have been made visually. An observer will compare the brightness of a target star with (if available) two other stars, one brighter and one fainter than the target. Experienced observers can often estimate magnitudes to ± 0.2 mag by this method.

Prior to the use of CCDs, photoelectric photometers were used by some amateurs to obtain high-accuracy light curves of bright variable stars. However, CCDs are far more sensitive and can be used to obtain light curves accurate to within ±0.05 mag of the true magnitude of the star being measured. Getting to these accuracies is not a trivial matter. It is true to say that most budding photometrists are daunted by the steps necessary to achieve this level of precision.

It's best to aim for a gradual progression to excellence rather than try to achieve world-class photometric results in one go. Software packages such as Richard Berry's *AIP*, *Maxim*, and *CCDSoft* can all plot the light curves, after analyzing multiple frames of a variable star. They will also lead you through the photometric process step by step; *AIP* is especially good in this regard. Lesser packages are to be avoided unless you have categorical proof that they work! Various early software routines offered photometric packages that were far from accurate.

Two specific questions which need to be addressed at the outset are the CCDs linearity, and its sensitivity across the spectrum. CCDs with antiblooming gates become increasingly non-linear as the CCD approaches saturation, so exposures need to be kept short such that the vital stars do not cause any pixels to exceed 50 percent saturation. In addition, professional photometry in the visual band requires a *V* filter (see the section on Photometric Filters, below) to ensure that all observers are measuring light across the same spectral band; put more simplistically, star magnitudes need to be measured at the same color. This does not mean that unfiltered observations are worthless as most of the time an amateur's unfiltered CCD magnitude estimate will be no less accurate than the best amateurs' visual estimates, especially where stars fainter than mag 14 are being measured. However, it is essential in all variable-star work that the comparison stars are noted, and the CCD/filter used is noted too, This enables "like to be compared with like."

CCDs are far more sensitive in the red/near-infrared than the human eye. This means that if the target or comparison stars emit strongly in the far red, the magnitude estimate will be different from a true *V*-band estimate. Specifically, unfiltered CCDs can suffer problems when stars may show large fluctuations in brightness in a narrow band but remain relatively unchanged across the wider spectrum. An example of this was the unusual nova discovered in Cassiopeia in December 1993 by Kanatsu. This nova exhibited a *DQ Her* type fade due to dust emission blocking the light output. As longer wavelength, near-infrared light penetrates dust far more easily than the shorter, blue wavelengths, the CCD magnitude of this nova was far brighter than the visual magnitude during the fade period.

Regarding the non-linearity aspect, it is essential to determine by experiment just how linear (or not) your particular camera is. Many CCD cameras come with software packages that offer a photometric option and even give an impressive two-decimal-place magnitude comparison when you click on a star. However, such a measurement is almost meaningless unless the manufacturers have taken the camera's linearity into account *and* a photometric filter is used. And if the star being measured is faint and only just above the noise floor, the uncertainty in the magnitude will be considerable, e.g., mag 16 ± 1 magnitude (the star is somewhere between mag 15 and mag 17!).

Professional astronomers with fully calibrated and filtered CCD systems can achieve photometric precision of 0.01 magnitudes with a single comparison star in the field; but even advanced amateurs struggle to achieve 0.05 magnitude precision. It is useful to remember that as a rough guide, 0.01 mag = 1 percent and 0.1 mag = 10 percent.

Unless you are sure of the linearity of your particular system, the safest plan is to do what most visual observers usually do – use two reliable comparison stars in the field.

Unfortunately, the photometric accuracies of the stars in the *Hubble Guide* star catalog are *not* reliable; in some cases they are one or two magnitudes in error. A star field containing photometric sequence stars should always be used where possible. Fortunately, many of the well-observed variable stars have reliable photometric sequences. The US Naval Observatory (http://www.usno.navy.mil) UCAC 2 or A2.0 catalogues can prove invaluable in this respect.

If you use two comparison stars they should, ideally, be just above and just below the target star's magnitude, as this will allow accurate calibration of the camera's linearity for a given exposure. If your camera software allows you to monitor the digital output from the A/D converter (prior to any software "fixes" to correct camera anomalies) you can be more confident that you are aware of your system's own deficiencies. It is also absolutely imperative that no image-processing routines (except basic dark-frame subtraction and flat-fielding) are carried out prior to the photometric analysis.

Photometry to even greater precision is possible with meticulous care and with the sort of software packages I have already mentioned, for example *AIP*, *Maxim* and *CCDSoft*.

The choice of aperture is another factor. In this context I am not talking about telescope aperture but the area on the CCD chip from which the charge is being recorded. We want to measure all of the light coming from each star, until it fades into the background sky brightness, but we don't want to include the light from any faint stars that may be lurking near the star in question. Of course, we also have to measure a blank area of sky of the same aperture so we know how much of the light in each aperture comes from the background sky. In practice, the optimum aperture to use is a circle of diameter 10–40 arc-seconds, with apertures above 20 arc-seconds being preferred when seeing and/or telescope tracking are poor. At 2 arc-seconds per pixel a typical aperture diameter might therefore be 10 pixels.

Photometric Filters

Professional astronomers measure magnitudes in accordance with internationally agreed standards – the sensitivity of their equipment is accurately defined at specific wavelengths. A series of pass-bands has been defined from the ultraviolet through to the infrared region and the most widely used system for defining these bands is the Kron–Cousins *UBVRI* system. (U = Ultraviolet; B = Blue; V = Visual; R = Red; I = Infrared).

Predictably, the *V*-band is the band of interest for most amateurs. It approximates the visual band of the human eye so that CCD magnitudes through a *V*-filter are directly comparable with visual magnitude estimates. But the *B*- and *R*-bands are also of interest. Some variable stars vary considerably in the blue end of the spectrum, even more than they do in the visual band. CCDs are at their most sensitive in the *R*-band, so this region is also of interest. The *I*-band is of interest to specialist professional astronomers but the *U*-band is close to the limit of most CCDs' spectral range.

If the magnitude of a star is determined in V and B pass-bands, then $B–V$ (B minus V) is the "color index." The color index can tell us a lot about the star; it can also be used to calibrate a photometric CCD system.

When choosing filters for a CCD camera it is important to understand that the Kron–Cousins band-pass boundaries define an ideal system and it is not possible to perfectly match a given set of filters to a CCD. The CCD will have its own response at specific wavelengths and to derive the response of the system it is necessary to *convolve* the spectral response of the chosen filters with the spectral response of the CCD chip. By convolve, we mean multiplying each point on the curve of the filters' spectral response, for every wavelength, with each point on the curve of the CCD's spectral response.

These days most CCD manufacturers will sell you, or recommend third-party vendors for, appropriate filter sets to match their CCD cameras. As an example, SBIG can sell you specific photometric filters for their CFW-8 filter wheel, or a complete photometric filter wheel matched to their camera's sensitivity.

It is worth remembering that a *V*-filter will easily knock 1.5 magnitudes or more off your CCD camera's magnitude limit. When you add the fact that you need a strong signal from the star (but not enough to take it beyond 50 percent saturation!) you need a surprisingly long exposure to get down to those 16th mag *V*-filtered stars that are beyond the visual observer – yet more proof of the formidable abilities of the human eye and brain combination, which can guesstimate a faint star's magnitude in an instant!

Specific Photometric Projects

Once you have a proven, working, photometric set-up, what observing projects are worth carrying out?

First of all, it is essential to contact the leading pro-am co-ordinators in this field. The AAVSO and the BAA have such co-ordinators; they can put you on an alert mailing list and advise you when an interesting or neglected CV is in outburst. Many professionals in the USA and the UK count on getting high-quality photometric data for their research projects, so you may suddenly become very popular!

Over the past few years Bill Worraker, in the UK, has organized a specific project to detect eclipses of "dwarf novae," another name for CVs which distinguishes them from the much rarer and more violent novae. For the record, a typical dwarf nova might be a CV that outbursts by 3–5 magnitudes every few months or even every few years, due to accretion disk instability. By comparison, a nova is a nuclear chain-reaction on the surface of a white dwarf, with an amplitude of 8–15 mags; these are generally one-off events, not outbursting again in a human lifetime (though there are recurrent novae).

Bill's project addresses something of a mystery in CV studies. Despite the huge numbers of CVs known and despite them all being binary systems, very few exhibit eclipses as seen from Earth. Obviously eclipses can only occur where the orbital plane of the system lines up with the Earth. This might seem unlikely, but remember, the two stars are *very* close together. When a dwarf nova outburst occurs and the accretion disk brightens to better than magnitude 15, there is an excellent opportunity for amateur astronomers to look out for the secondary star eclipsing the bright accretion disk. Out of over 400 dwarf novae which have well-documented outbursts, only a dozen or so show evidence of eclipses, despite the theoreticians predicting that over 30 percent should eclipse.

In theory if we are within ±20 degrees of the orbital plane of a dwarf nova, eclipses should be visible. Obviously theory and observation don't tie up here and

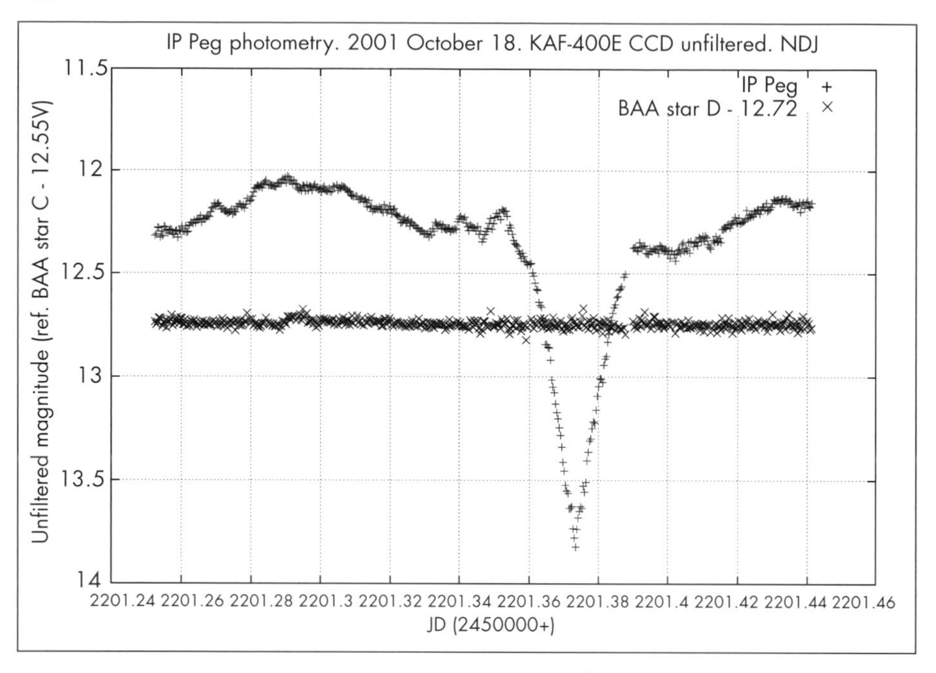

Figure 11.3. A dramatic light curve, by Nick James, of the cataclysmic variable IP Peg undergoing an eclipse, while in outburst. This was obtained with a 30-cm, f/5.25 Newtonian on 18 October 2001 over a 5-hour period. The straight line represents the photometry of a reference star of magnitude 12.72. Image: courtesy Nick James.

this is just one example of a field in which amateur photometrists are contributing vital scientific data.

A light curve (by Nick James) of the cataclysmic variable IP Peg undergoing an eclipse, while in outburst, is shown in Figure 11.3. This star can be as faint as magnitude 18, but reaches 12th magnitude in outburst. At this time, valuable photometry can be carried out even with amateur instruments. Nick uses a 30-cm Newtonian.

The eclipsing dwarf nova project is just one area where there is close cooperation between amateur and professional astronomers. Theoreticians have postulated that the inner parts of a dwarf nova's accretion disk may have mini-outbursts during quiescence. There is some evidence for this from light curves of the bright CV SS Cygni (the best-observed CV in the sky), but astronomers need to know if the outbursts are blue or red to find out if it is the secondary star or the accretion disk that is causing the outburst.

Amateur astronomer Karen Holland (Figure 11.4) is involved in this project and is using a special photometric filter wheel, provided by the BAA, for this work.

Photometry need not be restricted to variable stars either. By taking multiple images of close-approach asteroids, amateurs are now able (even without filters) to determine the rotation period(s) of asteroids which are usually rotating about

Figure 11.4. Karen Holland and her 25-cm LX200 Photo: courtesy Karen Holland.

one or multiple axes on time-scales of hours. If a bright asteroid is above the horizon in a clear, dark sky for several hours, some idea of its rotation period can be deduced from even one night of observations. More often, several nights of observations are combined, from various observers worldwide, to produce a complete light curve.

An example of an asteroid's light curves (obtained by Nick James) is shown in Figure 11.5. This shows the variation in the light output of asteroid 1998 WT24 which made a close pass by the Earth in December 2001. Nick's light curve shows a sinusoidal variation of about 0.25 magnitudes over a period of just over two hours.

Getting the light curve of an object which is flying along at almost a degree per hour is quite difficult, but, despite the problems, Nick managed to get two good runs on the object and acquired over 350 CCD frames on each occasion. Dr. Petr Pravec combined them with data obtained at the Ondrejov Observatory in the Czech Republic to derive a rotation period of just under 3.7 hours. This is consistent with radar results obtained by Steve Ostro's team using the Goldstone and Arecibo radio dishes.

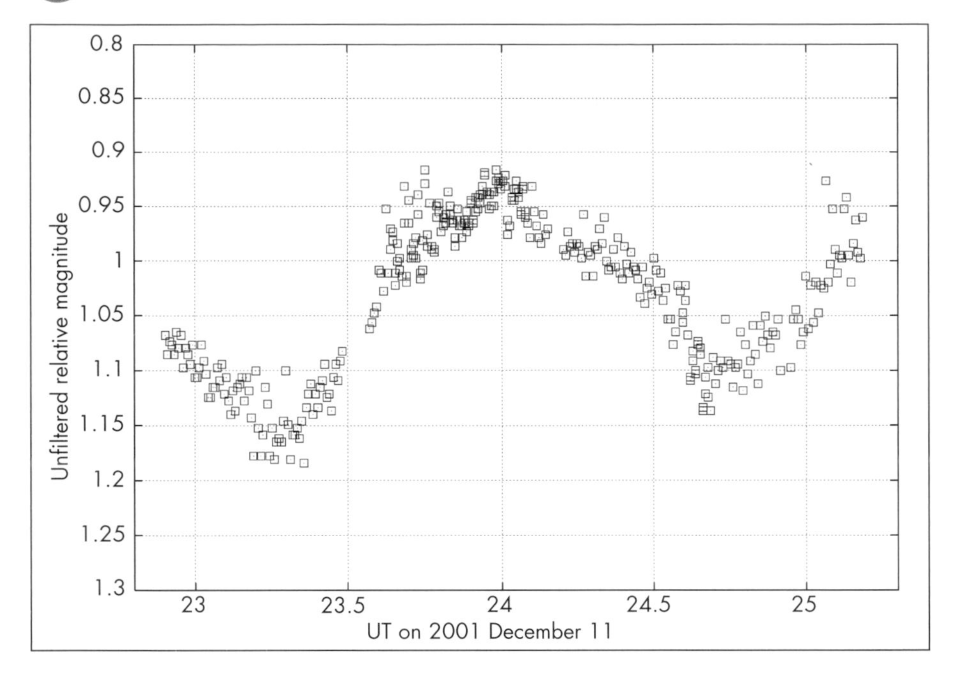

Figure 11.5. A light curve of asteroid 1998 WT24 secured by Nick James on 11 December 2001. When combined with professional results, a rotation period of just under 3.7 hours was derived.

Gamma Ray Bursters – The Ultimate Photometric Target!

Although not in the variable-star category, perhaps the rarest (but most highly prized) photometric observation an amateur can make is to confirm the optical counterpart of a Gamma Ray Burst (GRB). GRBs are the most powerful events known. In a period of seconds or minutes the burst can exceed the total gamma-ray output of the rest of the Universe as seen from Earth. Within minutes or hours, the burst can be gone! A few optical counterparts have been associated with GRBs, but they fade rapidly too; 24 hours after a GRB you typically need *Hubble* or the Keck to detect the afterglow, if it is visible at all.

Originally it was thought that GRBs were associated with neutron stars in our own galaxy. This was convenient, as the energy output would not have to be extraordinary if they were (relatively) nearby. However, the *Compton* satellite's BATSE detector (launched in 1991) showed that these objects are not distributed in the galactic plane. They are far beyond our galaxy, meaning that their power output is colossal.

It was soon realized that there was merit in trying to arrange for large telescopes to slew to the rough position of a GRB, within minutes of a discovery. This might reveal a fading optical counterpart. However, there were (and still are) two problems. First, GRB positions are frequently only known to an accuracy of 10 or 20 arc-minutes, a wide area for the narrow field of a big telescope's CCD. Second, professional telescopes are in permanent use and cannot always be diverted from their allotted program – and they may be in daylight when a GRB goes off. This is why amateurs have been enlisted to help detect GRB optical counterparts.

There is actually a case of a GRB reaching an optical magnitude of 9! Yes, it would have been visible in binoculars – on 23 January 1999, a BATSE alert enabled a robotic CCD detector attached to a 400-mm lens to image the 9th mag GRB within seconds of the outburst. This is very unusual and in the vast majority of cases amateurs need a capability to image down to at least mag 18 to even *consider* taking part.

There now exists (2003) a rapid alert network by which amateurs can be informed of GRBs by receiving SMS text messages on their mobile phones. For

a

Figure 11.6. a Arto Oksanen, Europe's leading GRB hunter and imager, and the Nyrola Observatory's 16-inch (0.4-m) Meade LX200 at Jyväskylä, Finland. Photo: courtesy Arto Oksanen.

(Figure 11.6. b, see overleaf)

b

Figure 11.6. b A full view of the Nyrola Observatory's 16-inch (0.4-m) LX200 at Jyväskylä, Finland. Note the high angle of the polar axis at latitude 62 north. Photo: courtesy Arto Oksanen.

more information on this check out the AAVSO Web page or *The Astronomer* magazine. Web pages for both are listed in the Appendix.

Since early 2003, the *Integral* satellite has provided a steady flow of GRB events for amateurs and professionals to follow.

Amateurs have already detected fading GRB optical counterparts. One of the most celebrated cases, certainly in Europe, was the detection of the optical counterpart of GRB 000926 on 28 September 2000 by the Finnish observer Arto Oksanen at Nyrola Observatory (Figure 11.6a, b). In response to an alert from professional astronomers Arto used a 0.4-m Meade LX200 operating at f/6.3 to capture 20 images of 4 minutes duration. When these images were stacked together and scrutinized by Hitoshi Yamaoka, Kyushu University, Japan, and later by Guy Hurst of *The Astronomer*, there was no doubt that Arto had

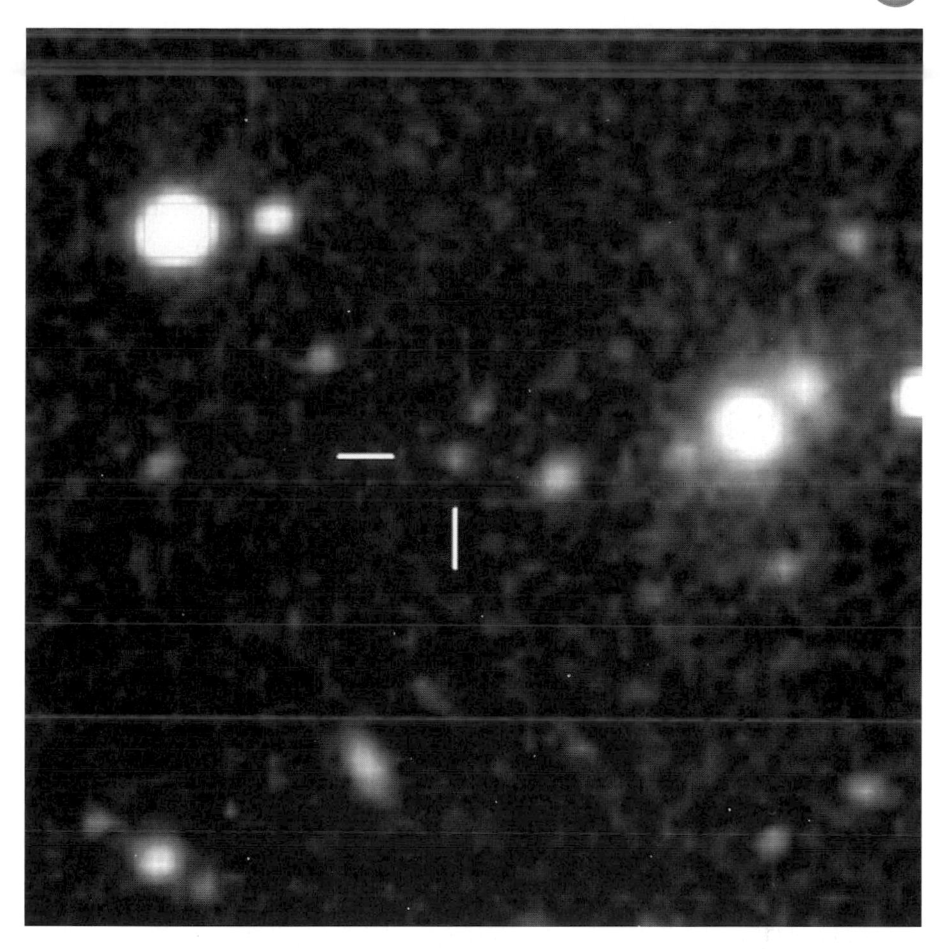

Figure 11.7. Arto Oksanen's image of the GRB which flared up on 26 September 2000. Imaged on 28 September 2002 from 18:18 to 19:40 UT with the 16-inch Meade LX200 of Nyrölä Observatory about 42 hours after the GRB erupted. SBIG ST7E unfiltered CCD camera. 20 × 240 second exposures with AO7 autoguiding. The GRB had faded to mag 20 at the time of the exposure. Photo: courtesy Arto Oksanen.

recorded the afterglow of the GRB (see Figure 11.7) at approximately magnitude 20! This was a remarkable achievement and further proof, if any were needed, that amateurs can equal professional observers even at the leading edge of modern astronomy.

On 4 October 2002 (as I started writing this chapter) I and half a dozen other amateur astronomers became the first UK astronomers to image the fading

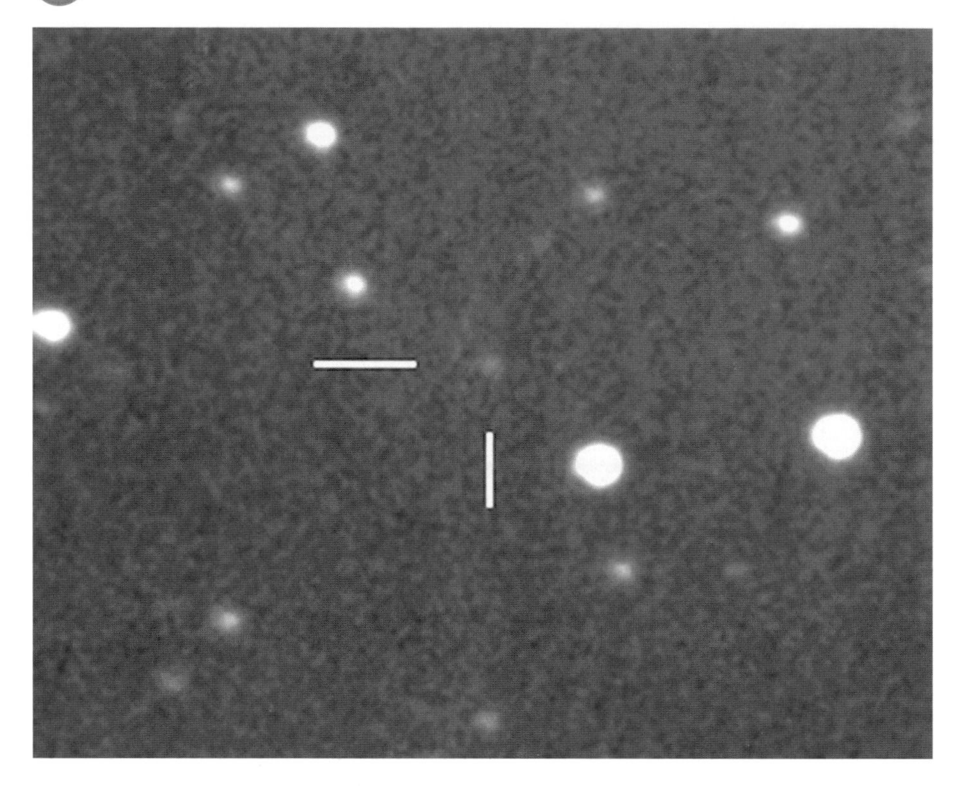

Figure 11.8. On 4 October 2002 the author and half a dozen other amateur astronomers became the first UK astronomers to image the fading optical counterpart of a gamma ray burst. The optical counterpart was magnitude 18 when the author secured this image with a 30-cm LX200 working at f/3.3. 4 October 2002. 20 × 100 second exposures, centered on 1945 UT. SBIG ST7 CCD.

optical counterpart of a gamma ray burst. The other amateurs were Nick James, Mark Armstrong, Tom Boles, David Strange, Peter Birtwhistle, and Eddie Guscott. My image is shown in Figure 11.8.

CHAPTER TWELVE

Saving the World: Near-Earth Object Chasers

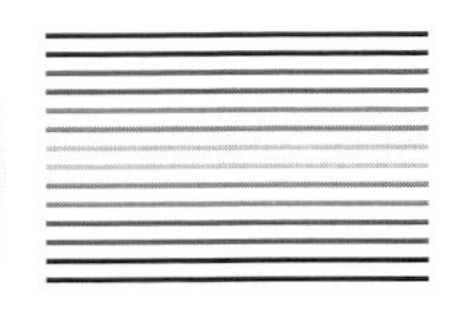

The Astrometrist: Saving the Earth from Destruction?

Before the era of CCDs, measuring the positions of new asteroids and comets was a tedious business involving the careful inspection of delicate negatives and time-consuming measurements with a mechanical "measuring engine." Now, with the appropriate software and a few clicks of the mouse, a precise measurement of the object can be determined with reference to dozens of background stars from the *Hubble Guide* star catalogue.

With this much power it might be thought that there was little left for astrometrists (astronomical measurers) to do. Indeed, in the case of bright new comets it is true that there is usually no shortage of astrometry. However, as automated discovery machines like LINEAR (the Lincoln Laboratories Near-Earth Asteroid Research telescopes), NEAT (Near-Earth Asteroid Tracking facility), and LONEOS (Lowell Observatory Near-Earth Object Survey) continue to trawl in thousands of asteroid discoveries, there is an increasing need to keep track of all these objects, especially the ones that are classed as NEOs – Near-Earth Objects – or PHAs – Potentially Hazardous Asteroids.

Potentially Hazardous Asteroids

I'm going to digress and touch on the subject of PHAs before returning to the actual astrometry issues later. A PHA is defined as any object of 150 meters or larger in size, with a minimum Earth orbit intersection distance of less than 7.5 million kilometers (0.05 AU). The Minor Planet Center (MPC) has a list on its Web pages of PHAs which will approach the Earth itself (as opposed to the Earth's orbit) within 0.05 AU in the next 176 years (see the Appendix).

There always seems to be "media hype" whenever a new asteroid is discovered which may, one day, come very close to the Earth. This publicity occurs despite the fact that new PHAs are discovered on a regular basis and despite the fact that it is impossible to tell whether an object will *actually* hit the Earth until a large number of astrometric measurements have been made by amateurs and professionals. More than 500 PHAs are now known and the number that satisfy the MPC criteria is increasing at the rate of more than one a week!

Richard P. Binzel of MIT has created a scale, called the Torino Scale, which grades the danger of a PHA from 0 (negligible risk) to 10 (global climatic catastrophe). Events graded as 10 (i.e., the certain impact of an asteroid of 150 meters or larger) may occur as regularly as every 100,000 years. The scale was named "Torino" because it was adopted at a conference in Torino (Turin), Italy in 1999. At the time of writing, the asteroid 2000 WO107 is predicted to pass closer to the Earth than any other known PHA in the next 176 years (i.e., up to 2178). On 1 December 2140 it will pass 80,000 kilometers from the Earth, little more than 6 Earth diameters away!

This object, discovered on 29 November 2000 by LINEAR, has an absolute magnitude (brightness when 149 million kilometers from both Sun and Earth) of 19.4, which corresponds to a diameter of approximately 500 meters. As well as PHA, the terms NEO (Near-Earth Object) and NEA (Near-Earth Asteroid) are sometimes encountered.

The terms NEO and NEA are generally used when an object is first discovered and little is known about its future hazard potential. NEO can also be a term applied to comets that may pass close to Earth. The risk from a comet impact is smaller, in as much as there are very few comets that come close to Earth. The short-period ones that do are often associated with meteor showers, like Swift–Tuttle (Perseids) or Tempel–Tuttle (Leonids).

A bigger worry, with comets that come in from deep space, is that from discovery to arrival at Earth there might be less than a year to prepare – not much of a warning! And big comets can have nuclei that are many kilometers (or tens of kilometers) across, so the impact when they hit Earth, at typically tens of kilometers per second, is potentially catastrophic for all life on Earth.

At present, estimates vary as to how many PHAs exist. There seems to be little slowdown in the rate these are being discovered, but it is essential to discover as many as possible so their orbits can be calculated and so we can ultimately catalogue all of the large ones. PHAs come in three categories determined by their orbital type (see Figure 12.1).

Firstly, the *Aten* category consists of bodies that spend most of their time within the Earth's orbit (i.e., close to the Sun), many of them crossing the Earth's orbit and thus posing a threat. The *Apollo* category consists of Earth-orbit-crossing bodies that spend most of their time outside the Earth's orbit. Finally, the *Amors* approach the Earth's orbit, lie within the orbit of Mars, and have their perihelia (closest point to the Sun) less than 1.3 AU from the Sun.

The Earth's atmosphere would only stop, or break-up, the smallest of these asteroids, those under about 50 or 60 meters across. Anything bigger (especially

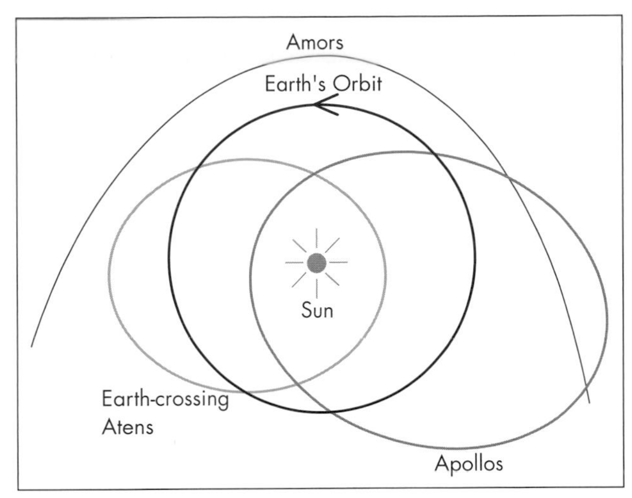

Figure 12.1. The orbits of Aten, Apollo, and Amor asteroids.

anything more than a few hundred meters across) would cause severe devastation whether it lands on solid ground and causes a "nuclear winter" and a huge crater, or lands in the sea and causes a tidal wave.

There may be a total of 2000 PHAs, in which case we have discovered no more than a quarter of them. In 1998 NASA stated that it aimed to discover 90 percent of all NEOs larger than 1 km, by 2010. At present the LINEAR facility in New Mexico is discovering more objects than NASA's NEAT telescopes at Palomar and Hawaii.

Whichever way you look at the problem, we simply don't know about most of the objects that are out there and which, one day, will certainly hit us.

Statistically, the chance of a big impact is very, very low in any one person's lifetime. However, if an impact occurs, the result could range from local devastation to a global catastrophe, wiping out all human life on the planet. As recently as 30 June 1908, a small comet or asteroid exploded in the skies above Tunguska, Siberia, laying waste to an area of 2000 square kilometers. This object may, originally, have been as small as 50 meters across, but it detonated with a force equivalent to 40 megatons. It devastated a tract of Siberian forest, but imagine the consequences if it had hit a major city. Then think about scaling that up for an object hundreds of meters across!

The Benson Prize

It is clear that not only discovering these objects, but accurately determining their orbits, is of vital importance. In 1997 James W. Benson of the Space Development Corporation announced the creation of the Benson Prize to reward the first 10 amateur astronomers who actually discover near-Earth asteroids.

The first winner of the award was US Arizona amateur astronomer Roy Tucker who has now discovered three near-Earth asteroids (1997 MW1, 1998 FG2, and 1998 HE3) and thus pocketed three $500 prizes. His discovery image of 1997

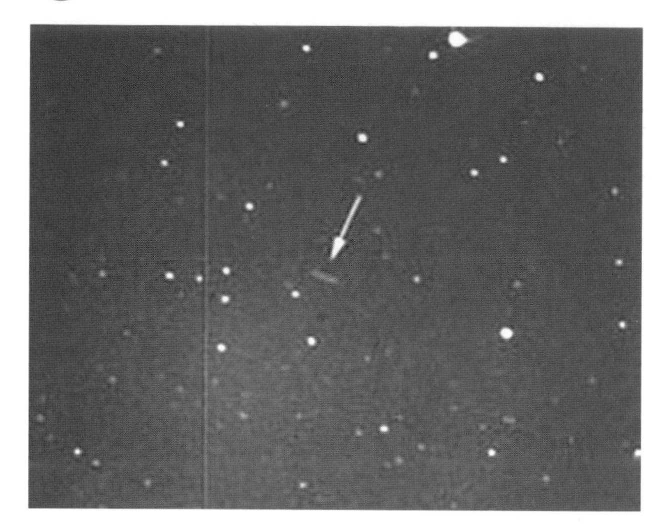

Figure 12.2. Near Earth Asteroid 1997 MW1 discovered by Roy Tucker. Photo: courtesy Roy Tucker.

MW1 is shown in Figure 12.2. Roy carried out a deliberate systematic search with a 14-inch (35-cm) Celestron SCT from his Tucson observatory. The technique he used was to employ a 1024 × 1024 pixel CCD camera on his SCT and then to take drift scan images of the sky. With this technique the telescope drive is turned off and the lines of the CCD are read out at the same rate that the stars move across it. With suitable software, a long strip of sky can then be assembled by combining many image downloads as the sky moves past the fixed telescope.

At the end of each scan, Roy "rewound" the telescope to the start point and started another scan. Finally, a third scan was completed. The next step was to blink all the images and search for any fast-moving objects. Although drift scanning results in a poor limiting magnitude, it is independent of mechanical drive errors (the telescope drive is off!) and sophisticated robotic control of the telescope is unnecessary for slewing to new fields. So if you don't have a Paramount ME or Astrophysics 1200GTO telescope mount, you can still search for asteroids!

Roy's original Celestron 14 system is shown in Figure 12.3a. He now has an array of three 35-cm reflectors employed in his new search for NEOs, also using the drift scan technique (see Figure 12.3b). Since 1998 there have only been two further amateur NEA discoverers. On 2 July 2000 school teacher Leonard Amburgey of Fitchburg, Massachusetts, discovered a 13th mag Apollo class asteroid 2000 NM while observing another asteroid in Serpens with his 212-mm aperture, f/3.9 Takahashi CN212 Newtonian. On 2 March 2002 the first Spanish NEA discovery was made. Amateur astronomer Rafael Ferrando was taking images of various comets (including the superb Ikeya–Zhang) before he moved on to the bright supernova 2002ap in M74. Later, when he slewed his 30-cm LX200 to the field of asteroid 2000 QW65 to carry out some astrometry, he noticed a faint streak on the image from his ST9 CCD camera. A second exposure confirmed that the object was moving rapidly. No such object was on the NEO Web page, so Rafael alerted the Minor Planet Center and continued taking images and carrying out astrometry. Other observers across the world were alerted rapidly (expect the MPC to respond very quickly to this sort of discovery – they always do!) and

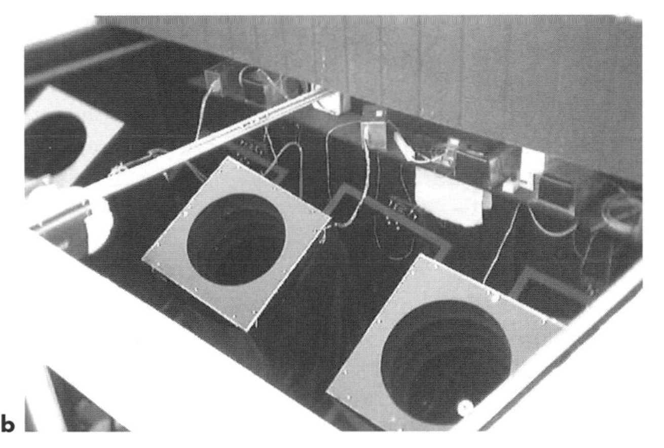

Figure 12.3. a Roy Tucker and his original Celestron 14 in its dome; this was used for his early NEO discoveries. **b** Roy Tucker's new NEO discovery system, comprising three 14-inch telescopes. Photos: courtesy Roy Tucker.

Ondrejov Observatory in the Czech Republic started making measurements within a few hours of the discovery. The object received the designation 2002 EA and turned out to be an Apollo asteroid with a diameter of more than 100 meters. A couple of weeks later it passed within 10 million kilometers of the Earth.

Although amateur discoveries of NEAs are rare, amateur discoveries of asteroids are not. The amazing amateur T. Kobayashi of Oizumi Japan has over 1500 numbered asteroids to his credit and uses nothing more than a 0.25-m, f/4.4 reflector!

The fact that only one amateur, Roy Tucker, has discovered more than one NEA since the Benson prize was introduced shows how difficult it is to scour the skies for fast-moving objects. However, where amateurs can *really* perform is in astrometry – which is where I started this section before my digression into the discovery issue.

NEO Astrometry

The professional patrol telescopes operated under the LINEAR and NEAT programs net hundreds of new Atens, Apollos, and Amors each year. In 2001, 30 Atens, 200 Apollos, and 200 Amors were discovered. LINEAR discovered 22, 130, and 120 of these, respectively, and these were just the objects that received designations, not the ones that got away or were only seen on one night.

Needless to say, even with the ease with which CCD astrometry can be performed, the world's astrometric resources are stretched to keep up with this barrage of discoveries. But if there is any category of object for which astrometry is required it is this category; the fate of the world could depend on it!

The biggest shortfall in manpower is in the astrometry of faint objects (fainter than 18th mag). It is remarkable how quickly a small asteroid can fade to this sort of feeble brightness as it moves away from the Earth. An asteroid 400 meters in diameter may sound large (and it would look pretty large coming down your street) but, in astronomical terms, its absolute magnitude (149 million kilometers from both Earth and Sun) would only be about 20.

To get an accurate orbit it is necessary to get as long an arc of measurements as possible, so measuring positions as the object fades is vitally important. By their very "close-approach" nature many NEOs appear to travel rapidly across the night sky, so even if they are not faint, they will cross many CCD pixels per minute when being imaged; this reduces their detectability. While a modern CCD, attached to a 30-cm scope, is very good at getting down to mag 20 or below in exposures of 5–10 minutes, it won't register an object that is traveling across each pixel every few seconds.

Fortunately (if you can afford one) the Paramount ME tracks so accurately that it can be programmed to track on a fast-moving asteroid. The resultant image, in which the stars are streaked, will show the NEO as a point. Of course, if the NEO has only just been discovered, its precise motion may be uncertain; however, even setting the motion approximately will be a help. Modern astrometric software packages enable measurements to be easily performed on such images.

If you don't have a super-accurate robotic mount like the Paramount ME, Herbert Raab's software *Astrometrica* (see the Appendix) will enable you to stack short exposures while allowing for the object's motion. A star catalogue can be matched to one of the individual short exposures to enable easy identification of the field. The best catalogue to use is the USNO's A2.0 catalogue which is available on the Internet. Further details can be found at the Web URL listed in the Appendix.

So how do you find out which brand new discoveries are crying out for astrometry? The Minor Planet Center maintains a NEO confirmation Web page for newly discovered fast-moving objects and this page is frequently updated (often several times per day). If the objects listed there are too faint for your system there is another page labelled *Dates of Observations of NEOs not Seen Recently* which contains other objects badly in need of astrometry. Both these Web page addresses are listed in the Appendix.

Objects on the NEO confirmation page need measuring over a period of several hours throughout the night as this is vital to improve the orbit and enable recovery in subsequent days. Many of the objects on the list turn out to be genuine NEOs (asteroids or comets), some may not exist at all, and some may just be main-belt asteroids. The new objects that have only been observed on a single night are marked as such. Objects remain on the NEO confirmation page until there is enough orbital data to issue an MPEC (Minor Planet Electronic Circular). If objects remain unconfirmed after five days they are defined as "lost" – in fact some may never have existed: remember that some of these objects are extremely faint "one-night suspects".

Because it is vital to publish new astrometric data as soon as possible, amateur observers may see their measurements published at the highest professional level (i.e., by the MPC) within hours of the data being submitted. There are many fascinating Web pages about NEOs to whet the observer's appetite. The Near-Earth Objects Dynamics Web site "NEODyS" created by the University of Pisa has a lot of useful data on the most dangerous PHAs (see the Appendix for the URL). Their

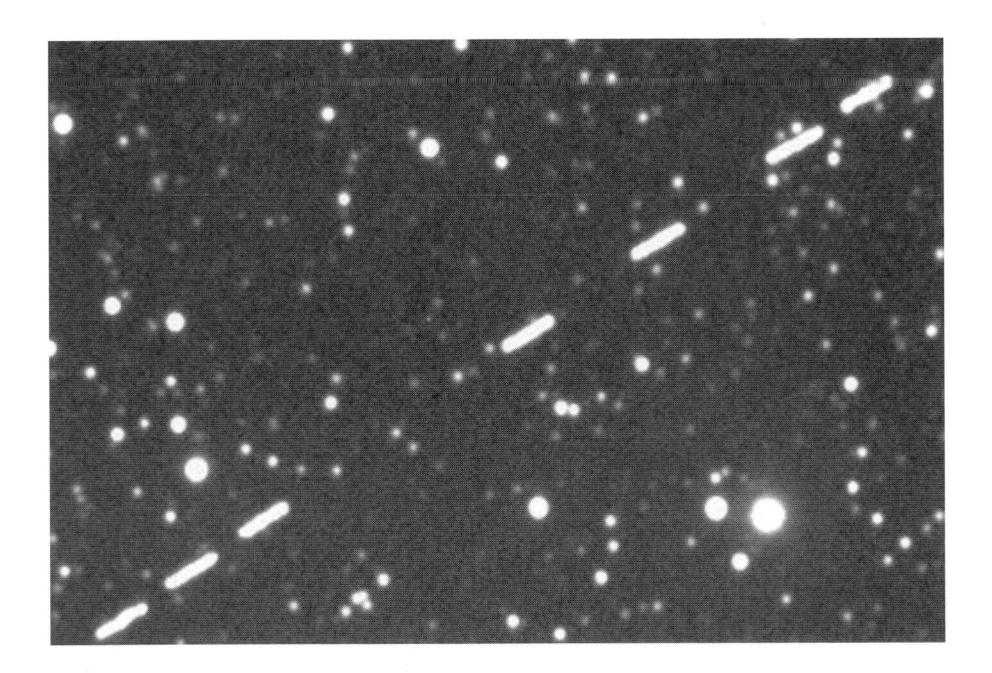

Figure 12.4. A composite, by the author of seven 60-second exposures, as Near-Earth Asteroid 1998 WT24 passed within two million kilometers of the Earth on 14 December 2001. The gaps are where poor telescope tracking resulted in an image being discarded. The field is 9 × 12 arc-minutes. 0.3-m LX200 and SBIG ST7 CCD. Photo: Copyright Martin Mobberley.

risk page lists all the objects which have a very small, but definitely non-zero chance of hitting the Earth, before 2080. Note that these are all objects which have been lost or have had insufficient astrometry to prove they will miss the Earth.

No objects yet discovered, and for which there is an accurate orbit, pose a serious threat in the next two hundred years.

So who are these amateur astronomers who carry out this vital NEO astrometry? Typical is Stephen Laurie, discoverer of supernova 1997bq in NGC 3147, and of 30 numbered asteroids between 1995 and 1997. In Stephen's supernova and asteroid hunting days he used a 25-cm LX200, but he now uses a far more powerful Celestron 14 (355-mm aperture) mounted on an Astrophysics AP-1200 mounting; this equipment can reach magnitude 20 in less than 10 minutes. With a focal reducer, his C14 has an f-ratio of 7.8, and the SBIG ST7E he uses gives a scale of 0.7 arc-seconds per pixel (9-micron pixels) with a field of view of 9 × 6 arc-minutes.

From Stephen's observing site (IAU Observatory Code 966), astrometry has been carried out on NEOs as faint as magnitude 20.7! In the June 2002 issue of *The Astronomer* magazine, Stephen reported his success rate with follow-up astrometry on objects listed on the NEO page. He obtained a success rate of less than 50 percent, mainly because the small field of view resulted in objects being missed. Despite this, only two months into his NEO observing program, the positions of 27 NEOs were obtained, by any measure a very impressive achievement!

Most NEOs are very faint, but occasionally a NEO whizzes close to the Earth and becomes very bright. My image of one such NEO is shown in Figure 12.4.

NEO astrometry is an exact science, requiring the most professional standards of observation and measurement. Despite this, many amateurs undertake it and – who knows – one day one of them may play an essential role in the survival of the human race.

CHAPTER THIRTEEN

Armchair Comet Hunters

Amateur astronomer Michael Oates (Figure 13.1) had no idea that comets could actually be *discovered* by browsing images posted on the Internet, until 29 January 2000, but incredibly, between 30 January 2000 and the end of July 2002, he discovered no fewer than *136* comets without ever going outdoors!

How is it possible that all these comets escaped detection by LINEAR and NEAT? Well, they are the Sun-grazing SOHO comets: tiny cometary fragments that only show up when they are a few solar radii out from the Sun and have been heated to extraordinary levels. They are called "SOHO comets", as opposed to being named after their discoverers, because they are discovered in the LASCO (Large Angle and Spectrometric Coronograph) images from the *SOHO* (Solar and Heliospheric Observatory) satellite.

Most of these comets are never observed with any other instrument. The *SOHO* satellite sits in a peculiar position where the gravitational pull from the Sun and Earth are the same; it is not a traditional "orbit" as such, just a stable position in space. LASCO looks at the Sun, but the dazzling solar disk is deliberately masked out. Because the instrument is in space there is no diffusion of light through the Earth's atmosphere, so the regions close to the Sun can be scrutinized for comets, just as you can do this during a solar eclipse.

To access the SOHO–LASCO images you need to go to the Web pages listed in the Appendix and follow the links from there. There is even a SOHO–LASCO Sungrazer page, also listed in the Appendix.

Many of the comet discoveries made by viewing the LASCO images have involved trawling through a huge number of archival images, so discovering comets while seated at a PC is not as easy as it may seem. There are also quite a few professional and amateur astronomers searching the images, so competition is intense! By August 2002, 500 SOHO comets had been discovered. The top

Figure 13.1. Michael Oates, searching for yet another SOHO comet! Photo: courtesy Michael Oates.

discoverers at that time were Michael Oates (136); Rainer Kracht (60); Xavier Leprette (58); Doug Biesecker (37); Michael Boschat (32); and Maik Meyer (28). Of these six, only Doug Biesecker is a professional astronomer.

Michael Oates uses *Maxim DL* to process the SOHO–LASCO images and a digital imaging software package called *ACDSee* to compare images and look for new comets. His determination has shown that the era of amateur comet discovery is *not* dead. SOHO comets can be discovered from indoors – you just don't get the satisfaction of having one named after you.

Figures 13.2a and b show two of Michael's SOHO comet discoveries. The SOHO–LASCO data used here are produced by a consortium of the Naval Research Laboratory (USA), Max-Planck-Institut für Aeronomie (Germany), Laboratoire d'Astronomie (France), and the University of Birmingham (UK). SOHO is a project of international cooperation between ESA and NASA.

Visual Comet Discovery and the Edgar Wilson Award

Although this book is primarily about the new "CCD era" amateur astronomers, I should mention the fact that *visual* discoveries of comets are

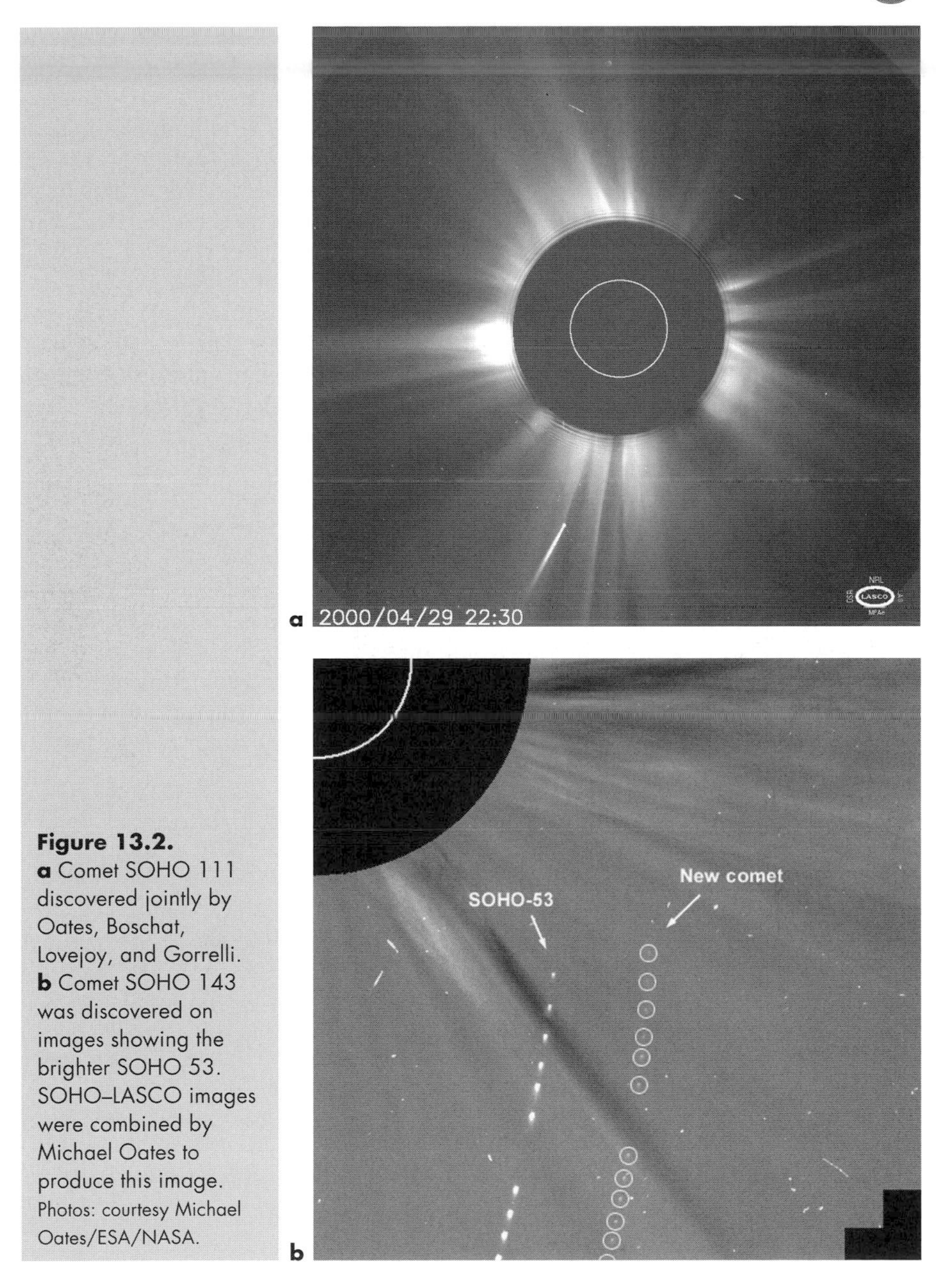

Figure 13.2.
a Comet SOHO 111 discovered jointly by Oates, Boschat, Lovejoy, and Gorrelli. **b** Comet SOHO 143 was discovered on images showing the brighter SOHO 53. SOHO–LASCO images were combined by Michael Oates to produce this image. Photos: courtesy Michael Oates/ESA/NASA.

still possible. As an incentive, the Edgar Wilson Award, administered by the Smithsonian Astrophysical Observatory, rewards amateurs who have discovered comets using amateur equipment (CCD discoveries by amateurs also classify).

A fund of $20,000 per annum is available. In July 2002 I wrote an article for *The Astronomer* magazine in which I described how five comets in recent years had avoided the discovery machines of LINEAR, NEAT, and LONEOS, to be discovered by amateurs. The comets in question were: C/2000 W1 (Utsunomiya–Jones); P/2001 Q2 (Petriew); C/2001 C1 (Ikeya–Zhang); C/2002 E2 (Snyder–Murakami); and C/2002 F1 (Utsunomiya). These prolific detection machines – incredibly so in the case of LINEAR – do have a few weaknesses. For a start, they are all based in the northern hemisphere and rarely discover anything further south than about –30 degrees Dec. Second, they avoid regions in the twilight or near the horizon, precisely where the amateur comet hunters hunt. Third, they rarely discover comets fainter than mag 19, so intrinsically faint comets that dart into twilight or the southern hemisphere as they brighten beyond this point, may be missed. Last, they also avoid the Moon and so may avoid parts of the sky for as much as two weeks; during this time a comet with a small perihelion, near to the Earth, can rapidly move into amateur detection range.

So although the machines will still get the vast majority of "potentially amateur-discoverable" comets, as much as a year before they enter the inner solar system, there is still hope for the visual comet hunter!

The diagram in Figure 13.3 shows how comet Ikeya–Zhang avoided LINEAR et al. in the year before discovery. If you assume that comets below mag 19, further south than –30 degrees and closer to the Sun than 90 degrees elongation, are out of LINEAR discovery range, the diagram instantly shows how this comet sneaked past.

In 2002, as well as the discoveries of the superb Ikeya–Zhang and comets Snyder–Murakami and Utsunomiya there were two other amateur comet discov-

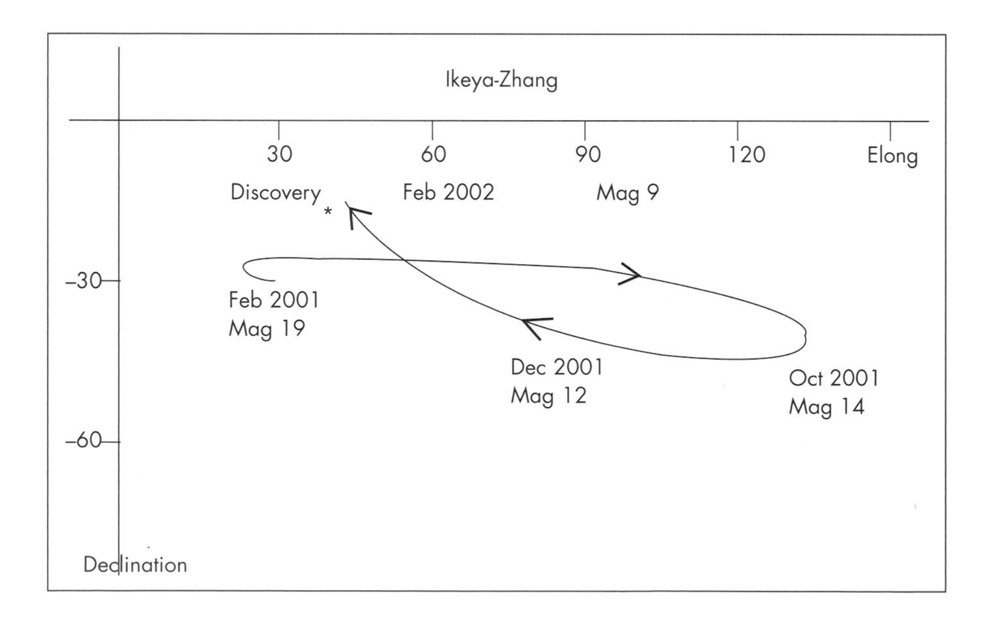

Figure 13.3. Ikeya–Zhang's track in the year prior to discovery. Diagram by the author.

eries of interest. On 21 January, William Kwong Yu Yeung of Benson, Arizona, discovered an object on a CCD image; this was originally classified as a 17th mag asteroid and designated 2002 BV. Subsequent observations by Timothy Spahr at Mount Hopkins, Arizona, revealed a coma and the object was reclassified as a periodic comet, P/2002 BV.

On 22 July 2002, Sebastian Hoenig of Germany discovered comet 2002 O4 while casually observing with a 25-cm LX200. Although already a discoverer of 11 SOHO–LASCO comets, Sebastian had spent five years hunting for comets visually. However, on this night he was casually observing deep-sky objects in the Andromeda/Pegasus region when he came across a new 12th mag comet that LINEAR and NEAT had missed; it was the first comet to have been discovered from Germany since 1946!

I have to say that I was especially pleased that the magnificent comet Ikeya–Zhang (see Figure 13.4) was discovered visually. Not only was it the best comet I had seen since Hyakutake and Hale–Bopp in 1996–1997, it was the longest-period comet (with a period of 340 years since the last return) that had definitely been observed twice. The previous record holder was Comet Herschel–Rigollet, with a 155-year period. When a historic comet like Ikeya–Zhang can be discovered visually, the last chapter on visual comet discovery has not yet been written.

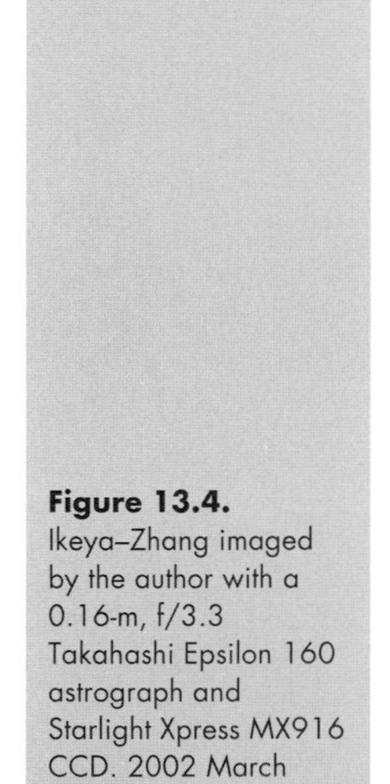

Figure 13.4. Ikeya–Zhang imaged by the author with a 0.16-m, f/3.3 Takahashi Epsilon 160 astrograph and Starlight Xpress MX916 CCD. 2002 March 26.822. 80-second exposure. Photo: Copyright Martin Mobberley.

CHAPTER FOURTEEN

Backyard Spectroscopists

Spectroscopy is not a field that has inspired many amateurs, despite the fact that professional astronomers have learnt more about the Universe from studying spectra than from anything else. Perhaps a long smear of gray-scales or color shades, crossed by vertical lines, just isn't as exciting as a beautiful image of Saturn or M51? Or perhaps most amateurs just don't want to get into a subject that involves complex optical equipment and an area of astronomy they simply don't understand? Most chapters on amateur spectroscopy that I have read begin by killing the interest of readers by immersing them in math, physics, and optics. That's why I'm going to leave that part until the end and start with the exciting things.

As everyone reading this book will surely know, a spectrograph splits the light from a star into a spectrum; if the spectrum is dispersed wide enough for useful resolution to be captured, it will be very faint, but a long CCD exposure can help compensate for this. Even the simplest prism or grating (I have seen net curtains used!) will split the light of the Sun into its constituent colors and show dark absorption lines that are produced by specific elements in the Sun's atmosphere. The extreme visual limits of the solar spectrum stretch from the H and K lines of calcium at 3934 and 3968 angstroms (one angstrom is 10^{-10} meters), deep in the violet, to the hydrogen-alpha 6563-angstrom line, deep in the red. Any useful spectroscope needs to be able to record details across this range.

SBIG's Spectroscope

Without a doubt the most exciting instrument in this field is Santa Barbara Instruments Group's (SBIG) $4000 spectrograph (Figure 14.1). SBIG's spectrograph

Figure 14.1. SBIG's autoguiding spectrograph attached to an SBIG ST7 CCD camera. Light travels from A (telescope interface) via a slit, to B (mirror) to C (collimating mirror) to D (the grating carousel.) The spectrum produced is then directed to the second half of the collimating mirror (E) which then focuses it into the CCD camera (G) via another mirror (F.) Photo: courtesy Maurice Gavin.

is designed exclusively for use with their autoguiding ST7 and ST8 CCD cameras, and with modern CCDs being three magnitudes more sensitive than the best photographic film, you'll understand that a modern 30-cm Schmidt- Cassegrain plus CCD will easily outperform a 1-meter instrument using film.

There is an exception – in the near-ultraviolet where film can rival older CCDs' performance, but newer CCD chips are pushing further into the blue end of the spectrum all the time. Perhaps equally important is the fact that CCD results can be analyzed as soon as the exposure ends; there's no dreary developing, fixing, and drying phase! So what science can be done with an instrument like the SBIG spectrograph?

SBIG's own promotional literature quotes a spectral signal-to-noise ratio of 10:1 for a 9th mag star with a 20-minute exposure using a non-ABG ST-7 CCD and a 25-cm aperture telescope in high spectral resolution mode (as good as 2.4 angstroms). The low spectral resolution mode (10 angstroms at best) will achieve the same signal-to-noise ratio with a mag 10.5 star using SBIG's narrow slit option. In practice, this specification tells us that spectra of bright galaxies (for example, the Messier galaxies) taken with the SBIG unit can easily show the red shift due to the expansion of the Universe, when compared with the spectrum of a nearby star.

Perhaps the most useful application of such an instrument in amateur hands is for the spectral monitoring of novae, which are usually discovered at mag 10 or brighter. Determining the spectral type of bright supernovae is also a possibility, although, even with CCDs, large amateur telescopes and long exposures are needed. In theory, a large (0.4–0.5 m) aperture amateur telescope should be able to take a suitable spectrum of a mag 15 supernova with an exposure of one hour; the signal-to-noise ratio would be poor, but good enough to discern the difference between a Type I or II supernova.

However, to my knowledge no amateurs seem to be doing regular supernova spectra work at the time of writing.

Spectroscopy Targets for Amateurs

In the UK, Maurice Gavin has been the leading pioneer of CCD spectroscopy for many years and has obtained numerous spectra of unusual variable stars and novae using a 30-cm Meade LX200 and home-made spectroscopes. For a few very

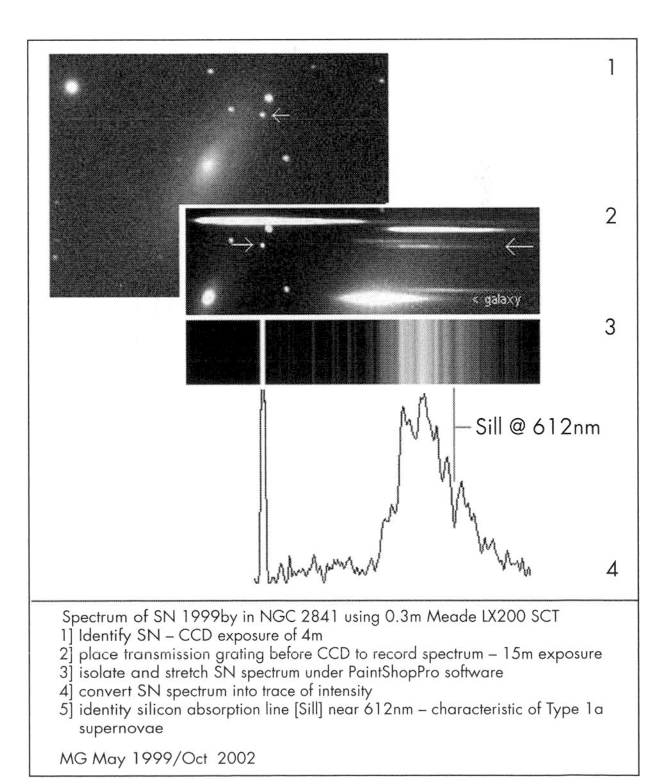

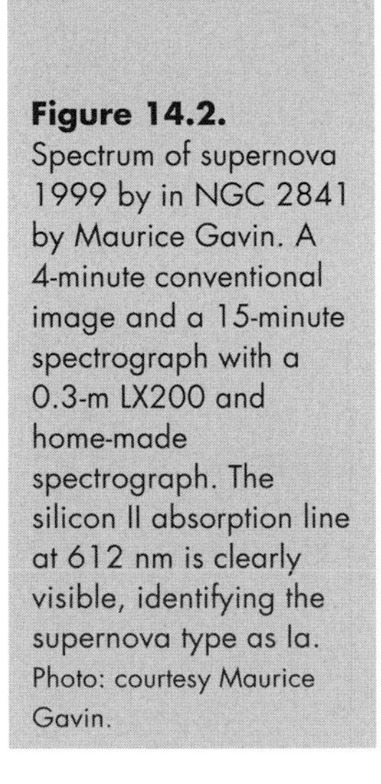

Figure 14.2. Spectrum of supernova 1999 by in NGC 2841 by Maurice Gavin. A 4-minute conventional image and a 15-minute spectrograph with a 0.3-m LX200 and home-made spectrograph. The silicon II absorption line at 612 nm is clearly visible, identifying the supernova type as Ia. Photo: courtesy Maurice Gavin.

bright (mag 12 or 13) supernovae, Maurice has attempted to determine the spectral type of the supernova.

Figure 14.2 shows his 15-minute exposure with a home-made spectrograph of supernova 1999by in NGC 2841. The silicon II absorption line at 612 nm is clearly visible, identifying the supernova type as Ia, a remarkable achievement.

Maurice has also taken spectra of numerous novae. An example of one such image is shown in Figure 14.3, of nova Cygni 2001 No.2. The H-alpha emission line, characteristic of novae, is clearly shown.

In the 1980s, amateur Gerald North used the 30-inch (0.76-meter) Coudé reflector at Herstmonceux in the UK (the site of the old Royal Greenwich Observatory – RGO) to take numerous spectra of the Moon, as part of the hunt for transient lunar phenomena (TLP: occasional colored glows and obscurations on the lunar surface). One of Gerald's aims was to emulate the spectrum obtained by the Russian Nikolai Kozyrev on 3 November 1958, using the 50-inch (1.27-meter) Crimea reflector. This spectrum allegedly showed molecular carbon vapor being emitted from the peak of the crater Alphonsus in a "volcanic process." This is highly controversial stuff and, to be frank, dismissed by many professionals as "ridiculous". However, actions speak louder than words and Gerald was determined to get as many spectra of the lunar surface as possible at the times when TLP were suspected by visual patrollers.

Sadly, this work is no longer being carried out, as the RGO no longer exists and no-one else has taken over Gerald's role. However, a modern CCD camera and spectroscope on a 30-cm reflector could easily duplicate Gerald's work with the 30-inch Coudé and with exposure times of only a few minutes. It would be nice to think that one day someone will work on this once again.

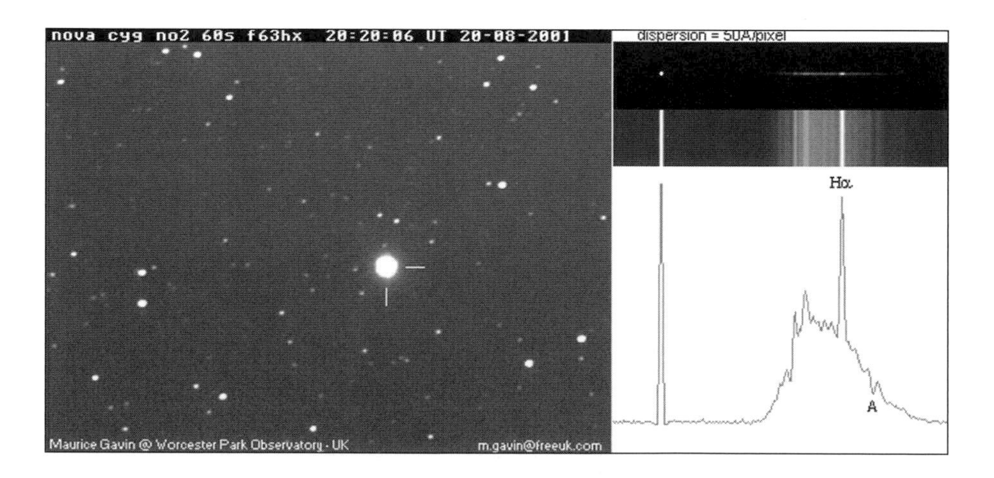

Figure 14.3. CCD image and spectrograph of nova Cygni 2001 No.2 by Maurice Gavin on 20 August 2001. The H-alpha emission line, characteristic of novae is clearly shown. 0.3-m LX200 and home-made spectrograph. Photo: courtesy Maurice Gavin.

How a Spectroscope Works

Okay, so now I've (hopefully) whetted the reader's appetite for spectroscopy, here comes the technical stuff!

The basic components of a single-prism spectrograph are shown in Figure 14.4. Essentially, the aim is to gather as much light as possible from the star being observed (and not from anything else), split the star's light up into a spectrum, and focus the spectrum. If a single narrow beam of light from an intensely bright point source (well, almost) like the Sun was being examined, all you would need is a chink in a curtain and a prism. But for astronomical spectroscopy with a telescope you need to channel parallel light from the star through the prism and then use a lens to focus the red end of the spectrum at one end of the CCD detector chip and the blue end at the other. This is the simplest, most efficient way to capture the spectrum.

Moving from left to right in the figure, we first come to the slit. In a normal telescope this is where the eyepiece would focus or the CCD would be placed – the focal plane, where the virtual image is formed. The purpose of the slit is to reduce background noise from the rest of the sky and to reduce overlap from adjacent wavelengths. The narrower the slit, the better the spectrum is resolved, *but* if the slit is narrower than the star image diameter at the focal plane, light will be lost. The collimator is simply a lens designed to ensure that the light rays entering the prism are parallel. Once the parallel light has been split into a spectrum by the prism, the spectrograph's own mini telescope lens – the imaging lens – easily focuses the red light on one end of the CCD and the blue on the other; so the spectrum is nicely spread out along the chip.

That's just the basics. In practice there are many variations. For example the prism can be replaced with a diffraction grating, which disperses the light in the same way. I'm ignoring a lot to try to keep things simple!

The next issues are how well can the spectrum be resolved, what focal length should the spectrograph telescope lens be, and how much of the spectrum will fit onto the length of the CCD?

With the typical prisms or gratings available to amateurs, the middle of the visual spectrum can be resolved as finely as one angstrom. But these same prisms

Figure 14.4. The basic components of a spectrograph. Diagram: courtesy of Prof. Chris Kitchin.

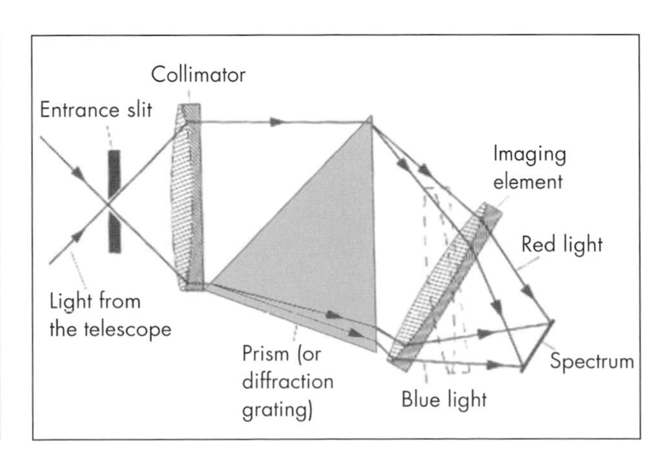

or gratings typically disperse the spectrum to such an extent that one angstrom of the spectrum subtends an angle of typically only 2 arc-seconds. This means that the spectrograph's imaging lens would need a focal length of a meter to capture one angstrom of resolution per 10-micron CCD pixel. At this scale, however, a 500-pixel long CCD array will only capture 500 angstroms of visual spectrum, compared to the whole visual spectrum of 4000–7000 angstroms, that is 3000 angstroms.

SBIGs unit features a choice of two dispersing diffraction gratings, offering 1 angstrom per pixel and 4.3 angstroms per pixel. But it also features an ingenious double concave mirror arrangement which acts both as collimator and imaging lens and keeps the unit's size compact. However, for the home-made spectrograph builder who wants to keep the imaging lens focal length short, settling for a resolution of a few angstroms per pixel and using a CCD with small pixels will help keep the system fairly small.

It's important not to confuse spectral resolution with dispersion. Let's look again at SBIG's spectrograph to clarify matters. The SBIG unit, like all spectroscopes, has a spectral resolution which is determined by the diffraction grating's performance, but this can be compromised if the slit is widened (to reduce exposure times) and by instrumental deficiencies. But to capture the resolution on the CCD, the dispersion and the focal length of the imaging lens/mirror must deliver a small enough "angstroms per pixel" scale. The SBIG unit features a choice of two diffraction gratings of 150 and 600 lines per mm with corresponding resolutions of 10 and 2.4 angstroms with the narrow, 18-micron slit (18 microns = 2 arc-seconds at 2 meters focal length). With the wide, 72-micron slit the 150 and 600 line gratings deliver resolutions of 38 and 10 angstroms. The dispersions of these gratings, combined with the focal length of the imaging lens/mirror, give image scales of 4.3 angstroms per pixel with the 150 grating and 1 angstrom per pixel with the 600 grating. The image scale is always of finer resolution than the spectral resolution to ensure that all the resolution available from the instrument is captured at the CCD.

Optimum grating/prism assemblies are rarely available to help the DIY spectroscope builder; likewise the collimating and imaging lenses. It's often a case of buying cheap components and bolting them together to see what happens! Amateur spectrographs are rarely designed precisely. Fortunately, diffraction gratings of 600 lines/mm can be purchased for as little as $25 and adjustable slits can be made from two razor blades. Second-hand camera lenses can be called into service for the collimating and imaging lenses, leaving the CCD as the most expensive component.

With a diffraction grating, dispersions are conveniently greater than with a single prism (older spectroscopes often use several prisms in sequence) but they produce two sets of spectra, each with several "orders" of spectra (see Figure 14.5).

The majority of the light goes into the white-light "zero-order" spectrum, but not all of it, so the spectra are not as bright as those made with a prism. However, if the grating is of the "blazed" type (more often found in reflection gratings), the individual grating line surfaces are angled to direct the majority of the light into the spectrum. To take advantage of this the grating has to be angled accurately to direct the bright spectrum at the detector.

Another issue is how to keep the telescope guided so that the star being analyzed is kept in the slit. One way of doing this is to focus a guiding eyepiece or

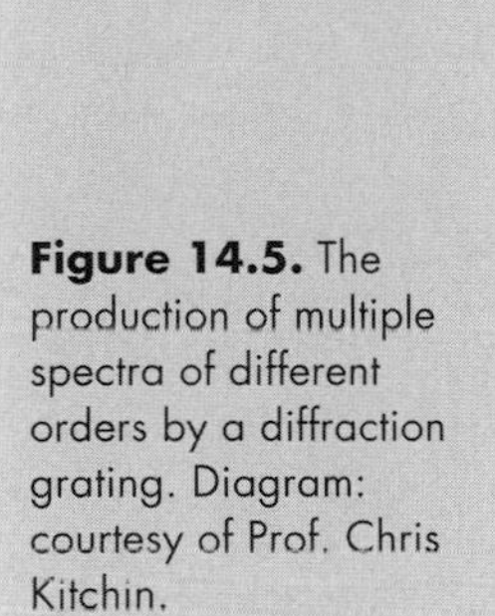

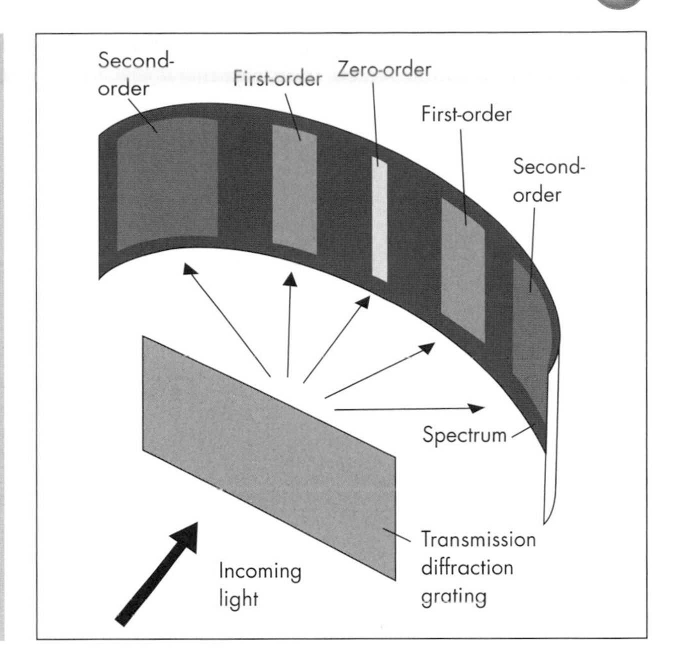

Figure 14.5. The production of multiple spectra of different orders by a diffraction grating. Diagram: courtesy of Prof. Chris Kitchin.

telescope on the outer surface of the slit; this surface, if highly polished, will easily show the outer overspill of the star's disk. It is actually advantageous to let the star's right ascension drift back and forth along the slit length as this produces the height of the spectrum. With perfect tracking the spectrum would be a thin line and very hard to analyze. The slit in the commercial SBIG unit is formed from two halves of a plane mirror which reflects the image at the focal plane to the separate guiding CCD. Thus, while the main CCD collects the spectra, the guiding CCD shows the field, with a dark line (or white if back-illuminated) showing the position of the slit; perhaps the ultimate spectroscope luxury!

Few amateurs will be able to spend $4000 on the SBIG spectrograph, but fortunately the basic components of a spectrograph are easily available. All that's needed is some engineering skill, lots of experimentation, and enough patience to fine-tune the instrument. A schematic showing the design of Maurice Gavin's

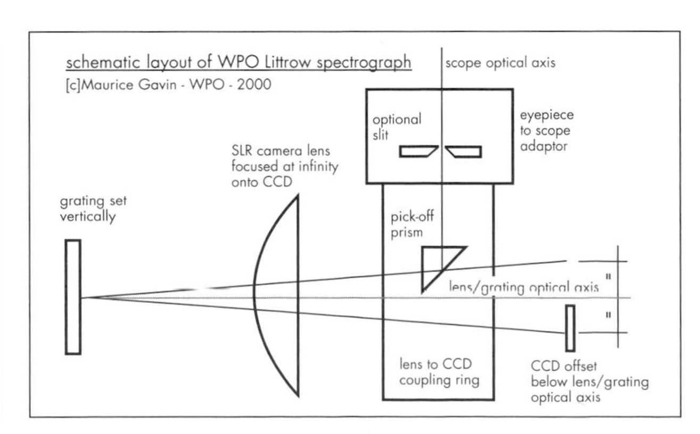

Figure 14.6. A schematic diagram of Maurice Gavin's Littrow spectrograph. Diagram: courtesy Maurice Gavin.

"Littrow" spectrograph is shown in Figure 14.6. His high-resolution spectrograph, attached to his LX200, is shown in Figure 14.7.

If you want more information on how to build a spectrograph, Maurice would very much like you to visit his Web page, the address of which can be found in the Appendix.

I hope that this short chapter will encourage others into the ranks of amateur spectroscopists, to obtain spectra of unusual variable stars, novae, and bright supernovae. This is real science, carried out in your own backyard.

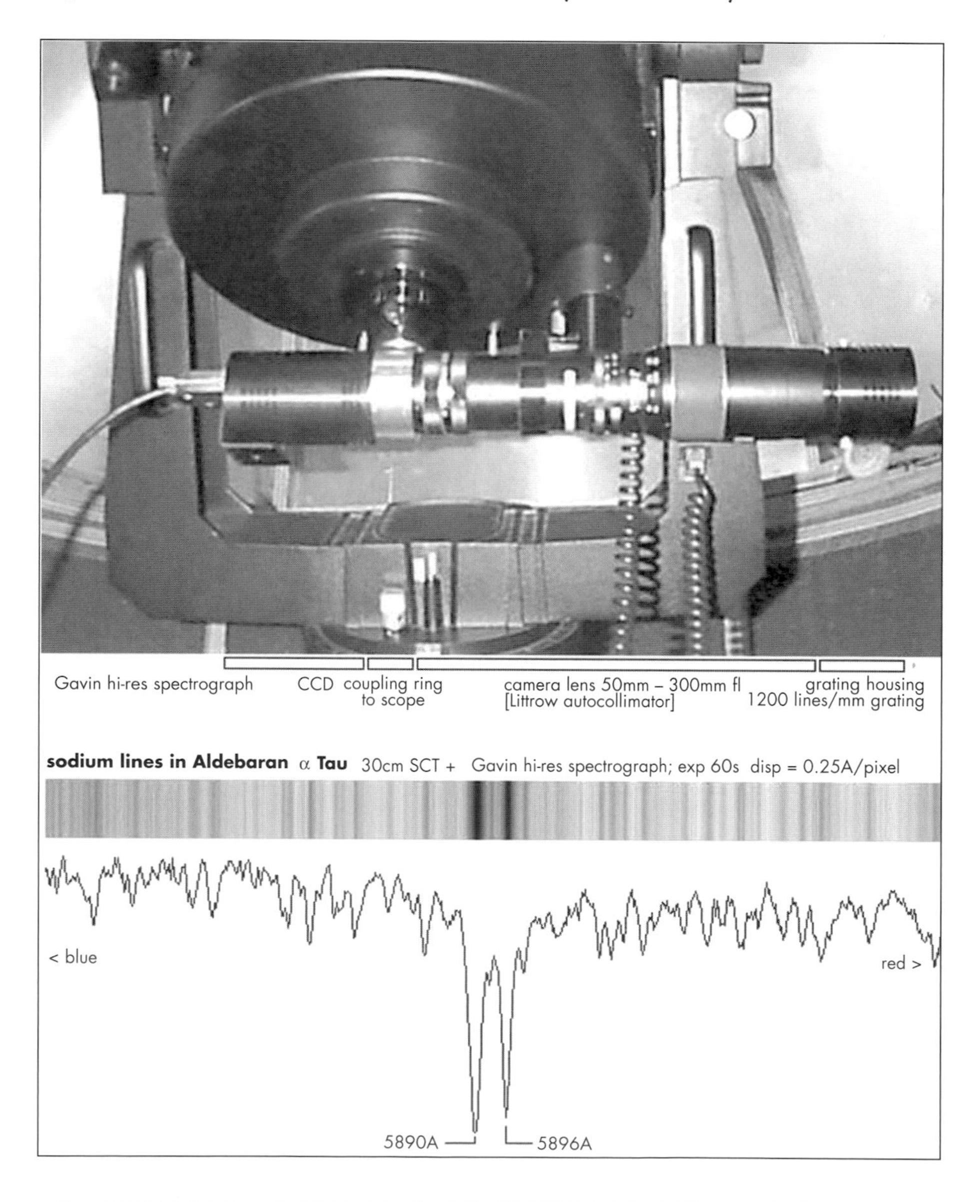

Figure 14.7. Maurice Gavin's high-resolution spectrograph attached to his 30-cm LX200. Diagram courtesy Maurice Gavin.

APPENDIX

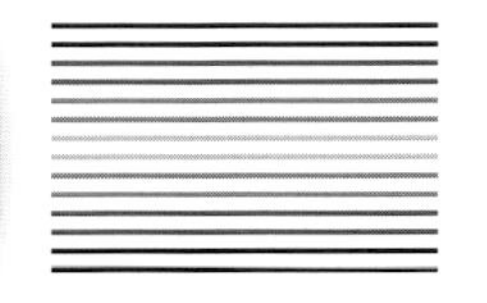

Useful Web Page URLs and Equipment Suppliers

Satellite Tracking

For information on *C-Sat* software for tracking the ISS and other satellites, go to:
www.skyshow.com

For information on the visibility of the International Space Station and other satellites go to:
www.heavens-above.com

Asteroid Occultations

In Europe, Ludek Vasta of the Czech Republic has set up an excellent EAON (European Asteroid Occultation Network) site at:
http://sorry.vse.cz/~ludek/mp/2001

The list of forthcoming events was completed by Edwin Goffin and was supplemented with the recommended observing window by Jan Manek. Edwin Goffin and Jan Manek's original Asteroid Occultation site with postscript files is in Belgium and is available via ftp at:
ftp://ftp.ster.kuleuven.ac.be/dist/vvs/asteroids.

Software for calculating asteroid occultation tracks (*Asteroid Pro*) is available:
http://www.anomalies.com/asteroid/info.htm

The world center for asteroid occultations, IOTA (International Occultation Timing Association) run by Dr David Dunham, is at:
http://www.anomalies.com/iotaweb/index.htm

Meteor Image Analysis

The International Meteor Organization's *Metrec* software Web site, containing the software of Sirko Molau, is at:
www.imo.net/video/meterc

Image Stacking

Robert J. Stekelenburg's *Astrostack* software can be found at www.astrostack.com

Webcam and Security Camera Astronomy User Group

Dr Steve Wainwright's Quick Cam and Unconventional Imaging Astronomy Group (QCUIAG) is a mine of information on converting webcams and security cameras for astronomical use:
http://www.astrabio.demon.co.uk/QCUIAG/

Suppliers of Low-Light Astronomy and Security Cameras

AVA, or Adirondack Video Astronomy, are pioneers in the supply of video equipment for Astronomy. Their address is 26 Graves Street, Glens Falls, NY 12801 USA.
They have a Web site at:
www.astrovid.com

AVA's UK dealers are True Technology Ltd, Wood Pecker Cottage, Red Lane, Aldermaston, Berks. RG7 4PA.
www.trutek-uk.com

In Europe the CCTV Model 2006X low-light camera is available from RF Concepts, Unit B2, Dundonald Enterprise Park, Carrowreagh Road, Dundonald. BT16 1QT N. Ireland. Web page:
http://www.rfconcepts.co.uk/

In the USA, the Watec 902H low-light camera is available from Rock House Products International, 2 Low Avenue, Suite 205, Middletown, New York 10940. Web page: http://rock2000.com/

The Supercircuits PC164C camera is available from Supercircuits at
One Supercircuits Plaza, Liberty Hill, Tx 78642. Web page: http://www.supercircuits.com

Supernova Suspect Checking

The CBAT/Minor Planet Center (http://cfa-www.harvard.edu/cfa/ps/cbat.html) has a supernova suspect check page at http://scully.harvard.edu/~cgi/CheckSN. This Web facility will tell you in seconds if there is an asteroid anywhere near your chosen galaxy.

Deep-Sky Survey

For photos of the sky down to mag 20, the Palomar DSS page is at: http://stdatu.stsci.edu/dss.

The Astronomer Magazine

The Astronomer Web site is at http://www.theastronomer.org and *The Astronomer* magazine editor, Guy Hurst, can be contacted at guy@tahq.demon.co.uk

Deep-Sky Excellence Links

Dr Kunihiko Okano of Japan, inventor of DDP, has a Web page at:
http://www.asahi-net.or.jp/~rt6k-okn/index.htm

The Kitt Peak Advanced Observing Program can be found at: http://www.noao.edu/outreach/aop.

The stunning gallery sections of SBIG are at:
http://www.sbig.com/sbwhtmls/gallery.htm

The stunning gallery pages of RC Optical Systems are at:
http://www.rcopticalsystems.com/gallery.html.

Variable Star Observing

The American Association of Variable Star Observers' (AAVSO) Web pages are at:
http://www.aavso.org/

The Variable Star Section of the British Astronomical Association is at:
http://www.britastro.org/

The Variable Star Network (VSNet) Web pages of Japan is at:
http://www.kusastro.kyoto-u.ac.jp/vsnet/

The Center for Backyard Astrophysics CBA Web pages (keen CCD photometry specialists) are at:
http://cba.phys.columbia.edu

NEO Astrometry Links

The MPC Close Approach Web pages, detailing objects which will approach the Earth itself (as opposed to the Earth's orbit) within 0.05 AU in the next 176 years, are at:
http://cfa-www.harvard.edu/iau/lists/PHACloseApp.html

Herbert Raab's software *Astrometrica* is available at:
http://www.bitnik.com/astrometrica/index.html)

The USNO's A2.0 catalogue is available on the Internet. Further details can be found at:
http://tdc-www.harvard.edu/software/catalogs/ua2.html.

The Minor Planet Center maintains a NEO confirmation Web page for newly discovered fast-moving objects at:
http://cfa-www.harvard.edu/iau/NEO/ToConfirm.html

The above page is frequently updated (often several times per day). If the objects listed there are too faint for your system there is another page labelled *Dates of Observations of NEOs not Seen Recently* at:
http://cfa-www.harvard.edu/iau/NEO/LastObsNEO.html

This contains other objects badly in need of astrometry.
The Near-Earth Objects Dynamics site "NEODyS", created by the University of Pisa, has a lot of useful data on the most dangerous PHAs at:
http://newton.dm.unipi.it/cgi-bin/neodys/neoibo

SOHO Comets

To access the SOHO–LASCO images to search for SOHO comets you need to go to:
http://sohowww.nascom.nasa.gov/data/realtime-images.html
or
http://lasco-www.nrl.navy.mil/lasco.html
and follow the links from there.

There is even a SOHO–LASCO Sungrazer page at:
http://sungrazer.nascom.nasa.gov/

Michael Oates uses *ACDSee* software to compare images. Their software is at:
http://www.acdsystems.com

Spectroscopy

Maurice Gavin's excellent Web pages on amateur spectroscopy are at:
http://www.astroman.fsnet.co.uk/begin.htm

Major Telescope Manufacturers

Meade: www.meade.com
UK dealer BC&F: www.telescopehouse.co.uk

Celestron: www.celestron.com
UK dealer David Hinds: www.dhinds.co.uk

Software Bisque (Software and the Paramount ME)
www.bisque.com

Astrophysics (the AP1200GTO mount and apochromat refractors)
www.astro-physics.com

Takahashi Dealers:
True Technology (UK) www.trutek-uk.com
Texas Nautical (USA) www.takahashiamerica.com

RC Optical Systems (RCOS)
www.rcopticalsystems.com

Optical Guidance Systems
www.opticalguidancesystems.com

Jims Mobile Inc. (superb focusers and accessories)
www.jimsmobile.com

TeleVue (eyepieces and small apochromat refractors)
www.televue.com

CCD Manufacturers

SBIG
www.sbig.com

Starlight Xpress
www.starlight-xpress.co.uk

Apogee
www.ccd.com

FLI
www.fli-cam.com

Audine (French CCD cameras in kit form)
http://www.astrosurf.com/audine/English/index_en.htm

Rockingham Instruments
www.rockinghaminstruments.com

Software

Software Bisque (*CCDSoft*, *The Sky*, *Orchestrate*, and *TPoint*)
www.bisque.com

Maxim (*MaximDL/CCD*)
www.cyanogen.com

AIP (The Handbook of Astronomical Image Processing)
Richard Berry's *AIP* book and CD can be ordered from Willman Bell. Published in 2000, ISBN 0-943396-67-0
www.willbell.com

AstroArt
www.msb-astroart.com

Project Pluto (*Guide 8.0*)
www.projectpluto.com

SkyMap Pro 8
www.skymap.com

Index

229